Geomathematics: Theory and Applications (Volume II)

Geomathematics: Theory and Applications (Volume II)

Virginia Williams

CLANRYE
INTERNATIONAL
www.clanryeinternational.com

Clanrye International,
750 Third Avenue, 9th Floor,
New York, NY 10017, USA

ISBN: 978-1-64726-658-5

Cataloging-in-Publication Data

Geomathematics : theory and applications (Volume II) / Virginia Williams.
    p. cm.
Includes bibliographical references and index.
ISBN 978-1-64726-658-5
1. Geology--Mathematics. 2. Mathematics. 3. Geology. I. Williams, Virginia.
QE33.2.M3 G46 2023
550.151--dc24

For information on all Clanrye International publications
visit our website at www.clanryeinternational.com

𝒞LANRYE
INTERNATIONAL

Contents

Preface

Geomathematics is the application of mathematical methods to solve problems in geosciences, including geology and geophysics. It is also called mathematical geophysics. There are several applications of geomathematics in different areas such as data assimilation, geophysics, terrestrial tomography, crystallography, geomorphology, glaciology and geophysical statistics. The field of geomorphology involves the application of geomathematics by using mathematical methods related to soil and water. It includes the application of various mathematical concepts such as Darcy's law, Stoke's law, stream power and differential equations. Seismic tomography, a technique used for the imaging of the subsurface of the Earth by using seismic waves, also makes extensive use of geomathematics. This book provides a detailed explanation of geomathematics. Its extensive content provides the readers with a thorough understanding of the theory and applications of this field. Coherent flow of topics, student-friendly language, and extensive use of examples make this book an invaluable source of knowledge.

This book has been the outcome of endless efforts put in by authors and researchers on various issues and topics within the field. The book is a comprehensive collection of significant researches that are addressed in a variety of chapters. It will surely enhance the knowledge of the field among readers across the globe.

It gives us an immense pleasure to thank our researchers and authors for their efforts to submit their piece of writing before the deadlines. Finally in the end, I would like to thank my family and colleagues who have been a great source of inspiration and support.

Virginia Williams

12

Study of Coast Lines Time Variations: A Morphological Perspective

Jean Serra

Abstract A morphological approach for studying coast lines time variations is proposed. It is based on interpolations and forecasts by means of weighted median sets, which allow to average the shorelines at different times. After a first translation invariant method, two variants are proposed. The first one enhances the space contrasts by multiplying the quench function, the other introduces homotopic constraints for preserving the topology of the shore (gulfs, islands).

Keywords Median sets · Binary interpolation · Hausdorff distances · Shoreline Time forecasting

12.1 Three Problems, One Theoretical Tool

The following study holds on lagoon inlets movements. It extends and develops an experimental study made by N.V. Thao and X. Chen about Thuan An Inlet Area (Thao and Chen 2005). The predictions proposed by these authors were obtained by averaging over the time the successive positions of a complex shoreline, including lagoon inlets, which results in a prediction of the coast line. J. Chaussard showed, in Chaussard (2006), that this prediction correctly fits with ulterior data from Google Earth (see Fig. 12.1).

In Thao and Chen (2005), the authors used a popular way to estimate accretions (Srivastava et al. 2005). Figure 12.2 depicts this semi-manual approach: the shoreline has been discretized into segments which are shifted upwards according a given accretion law (here the linear law $y = ax + b$, where x stands for the time). Indeed, this is nothing but a sampled version of the dilation the shoreline by the disc of radius $ax + b$. Such a circular dilation of a shoreline turns out to be the simplest expression of its evolution under an accretion process, since it is uniform everywhere and does not take the previous stages of the shoreline into account. As a matter of fact, the

J. Serra (✉)
Ecole des Mines de Paris, Paris, France
e-mail: jean.serra@cmm.ensmp.fr

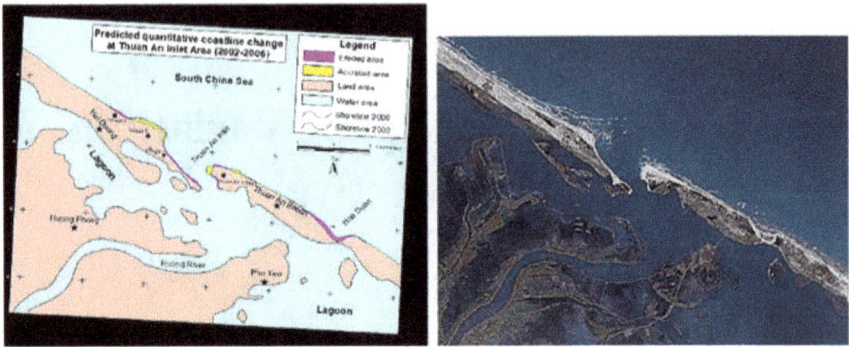

Fig. 12.1 Left: Lagoon Inlets forecast by N.V. Thao and X. Chen; right: Current Google earth view of the same area

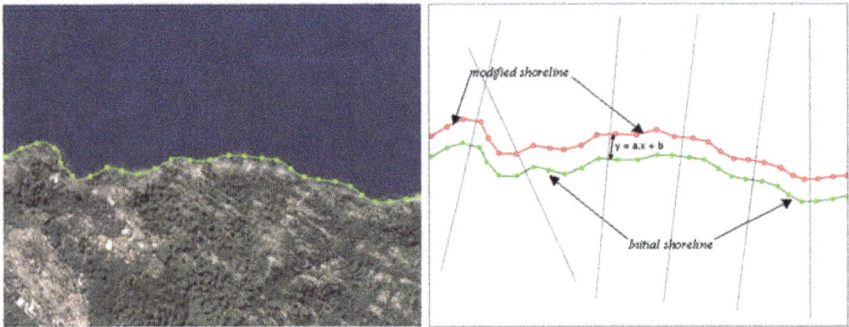

Fig. 12.2 Classical semi-manual technique of extrapolation

notion of a set extrapolation is not straightforward, and depends considerably on the features one wishes to preserve or to emphasize.

1. If, by comparing the shorelines at years n and $n-1$, there appear zones of erosion[1] and zones of accretion, we may require a forecast of the shoreline, at year $n+1$, to pursue erosions and accretions, but always in the same zones as previously; moreover, we must be able to express several laws for this time evolution (for example, in Thao and Chen 2005, a linear and a logarithmic laws are discussed);

2. if we know the movements of the shore during the last ten years, with one map per year, we can average these ten sets independently of their dates and base the extrapolation on this average only, or we can alternatively emphasize the more recent maps, considering that the last one, or the last two ones, carry most of the information;

---

[1] The shoreline context, the two words of "erosion" and "accretion" refer to the two types of changes depicted in Fig. 12.3. The word "erosion" also appears in the context of mathematical morphology, for naming the operation $\ominus$ involved in Eq. 12.1. It is pure coincidence.

3. if the shore exhibits small gulfs, islands and lagoon lakes, we may require from the extrapolation to preserve their homotopy, i.e. neither to create new islands (new gulfs, new lakes) nor to suppress the existing ones.

The first two questions can be treated within the framework of the median set theory, and the third one reduces to a small variant. Though median elements were thoroughly studied for interpolation problems, by M. Iwanowski in particular Iwanowski and Serra (2000) no attention was paid to their potentialities for generating averages and extrapolations. We believe nevertheless median sets turn out to be convenient tools for shorelines forecast, which in addition extend directly to numerical functions (however, we shall not treat the numerical extension here, and restrict ourself to the binary approach).

What follows is an attempt in this direction. After a presentation of the median set, that we adapt to shorelines in Sect. 12.2, we analyze in Sect. 12.3 a series of derived notions, such as weighted median set, quench function and quench stripe, and averages. The heart of the matter is treated in Sect. 12.4, where various laws are proposed for the dynamics of the coast movements. A short section on homotopy preservation precedes the conclusion. All images of coasts which are used below are *simulations*, and have the same digital size of 512×320 pixels.

12.2 Median Set

In literature, median set appears as an interpolation algorithm in Casas (1996) and in Meyer (1996), and was extended to partitions in Beucher (1998). Its formal definition and its basic properties were given in Serra (1998). Since, the approach has been developed by several authors (Angulo and Meyer 2009; Charpiat et al. 2006). In what follows, the geographical space is modelled by the Euclidean plane, but the approach applies as well to any metric space, including the digital ones. The model of Euclidean median sets does not concern the *lines* of the shores, but the *whole landsets*, whose the shorelines are the boundaries. These landsets, denoted below by A_1, A_2, etc., are depicted for example in Fig. 12.3 left, whereas the only shorelines boundaries, in another example, are depicted in Fig. 12.5 left. The basic results we need to start with are the Definition 1 of a median set, and the two properties 2 and 3, drawn from Serra (1998).

Hausdorff distance ρ concerns the class $\mathcal{H}'$ of the noncompact sets of R^n (here of R^2). It is the mapping $\rho : \mathcal{H}' \times \mathcal{H}' \rightarrow R_+$

$$\rho(X, Y) = \inf\{\lambda : X \subseteq Y \oplus \lambda B ; Y \subseteq X \oplus \lambda B\} \qquad (12.1)$$

where B designates the unit disc centered at the origin, and where $\oplus$ and $\ominus$ designate Minkowski addition (or dilation) and substraction (or erosion) respectively.

Consider now an *ordered* pair of closed sets $\{X, Y\}$, with $X \subseteq Y$, and such that the numerical value $\rho(X, Y)$, as given by Eq. (12.1), is finite. Their median element is defined as follows:

Definition 1 The median element between the two ordered sets $X, Y \in \mathcal{K}'$, with $X \subseteq Y$, is the compact set $M(X, Y)$, comprised between X and Y and whose boundary points are equidistant from X and Y^c.

In other words, the boundary ∂M of M is nothing but the skeleton by zone of influence, or *skiz*, between X and Y^c.

Proposition 1 *The median set between X and Y is obtained by taking the union*

$$M(X, Y) = \cup\{(X \oplus \lambda B) \cap (Y \ominus \lambda B) \ \lambda \geq 0\} \tag{12.2}$$

where the λ can be limited to the values smaller or equal to

$$\mu = \inf\{\lambda : \lambda \geq 0, X \oplus \lambda B \supseteq Y \ominus \lambda B\} \tag{12.3}$$

and where the equality is reached for at least one point of ∂M.

Proof A point m at a distance $\leq \lambda$ from X and $\geq \lambda$ from Y^c belongs to set $(X \oplus \lambda B) \cap (Y \ominus \lambda B)$, hence to set of Eq. (12.1). Conversely, as every point $m \in M$ belongs to at least one term of the union, there exists a $\lambda \geq 0$ with $d(m, X) \leq \lambda$ and $d(m, Y^c) \geq \lambda$, which results in Eq. (12.1). As for Eq. (12.2), we observe that for λ large enough we have $(X \oplus \lambda B) \cup (Y^c \oplus \lambda B) = R^2$ because set Y is bounded. These λ bring no contribution to set $M(X, Y)$, since $X \oplus \lambda B \supseteq Y \ominus \lambda B$. Finally, for $\lambda = \mu$, we obtain a point of the boundary ∂M because X and Y are closed, which achieves the proof.

Here is now an instructive property which shows how both Hausdorff distances by dilation and by erosion[2] are involved in the median $M(X, Y)$ (Serra 1998).

Proposition 2 *Given $X, Y \in \mathcal{K}'(R^n)$, the median element $M(X, Y)$ is at Hausdorff dilation distance μ from X and from the closing $X \bullet \mu B = (X \oplus \mu B) \ominus \mu B$, and at Hausdorff erosion distance μ from Y and from the opening $Y \circ \mu B = (Y \ominus \mu B) \oplus \mu B$.*

---

[2]Hausdorff distance σ for erosion, introduced in by the relation

$$\sigma(X, Y) = \inf\{\lambda : X \ominus B_\lambda \subseteq Y ; Y \ominus B_\lambda \subseteq X\}$$

concerns the subclass $\mathcal{A}$ of $\mathcal{K}'(E)$ of the regular compact sets, i.e. such that $\overline{X^o} = X$. It is indeed a distance on $\mathcal{A} \times \mathcal{A}$. If $\sigma(X, Y) = 0$, then we have

$$Y \supseteq \bigcup_{\lambda > 0} X \ominus B_\lambda = X^o \ \Rightarrow \ Y \supseteq \overline{X^o} = X \qquad X, Y \in \mathcal{A}$$

and similarly $X \supseteq Y$, hence $X = Y$ (the other two axioms are proved as for distance ρ) (Serra 1998).

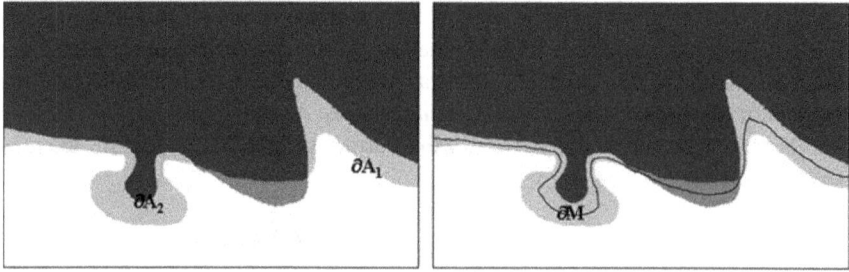

Fig. 12.3 Left: two simulated shore images A_1 and A_2. The older is supposed to be A_1 (the white one). The zones of accretion from A_1 to A_2 are in light grey, those of erosion in dark grey; right: the boundary of the median set M between A_1 and A_2

The Hausdorff distance applies to non empty compact sets. But clearly, the landsets under study are not empty, and the above assumption that $\rho(X, Y) < \infty$ comes back to say that all involved distances are bounded.

12.3 Median and Average for Non Ordered Sets

Non ordered sets In general, two successive shores A_1 and A_2 are not ordered, i.e. their change comprises both erosions and accretion areas. If so, the previous results do not apply to two A_1 and A_2 directly, but to their intersection $X = A_1 \cap A_2$ and their union $Y = A_1 \cup A_2$ which are ordered since $X \subseteq Y$. Equation (12.1) of the median element becomes

$$M(A_1, A_2) = \bigcup_{\lambda \geq 0} [A_1 \cap A_2) \oplus \lambda B] \cap [(A_1 \cup A_2) \ominus \lambda B] \qquad (12.4)$$

Figures 12.3 depicts an example of median set M. One observes that ∂M goes through all points where the two coastlines intersect. The property is general, since these points belong to both $A_1 \cap A_2$ and $A_1 \cup A_2$.

Weighted median Set M is said to be *median* because each point of ∂M is equidistant from X and Y^c, which is a consequence of the same weight given to dilation and erosion in Eq. (12.2). By changing this weight, i.e. by replacing M by

$$M_\alpha(X, Y) = \bigcup_\lambda \{(X \oplus \alpha \lambda B) \cap (Y \ominus (1 - \alpha)\lambda B)\} \qquad (12.5)$$

for a $\alpha \in [0, 1]$, we generate another interpolation, and by making α vary, a series of progressive interpolations from X to Y (Huttenlocher 1995), all the closer to set Y since α is high. One will notice that when the two shores A_1 and A_2 are *not* nested in each other, then one takes for the two operands of Eq. (12.5) $X = A_1 \cap A_2$ and

$Y = A_1 \cup A_2$. This provides interpolators such as those of Fig. 12.4. Unfortunately, these interpolators are closer to the highest or to the lowest line, no matter these lines are portions of ∂A_1 or of ∂A_2. For correcting this drawback, one must take the interpolator M_α in the zones where A_1 is larger than A_2 (for example), and $M_{1-\alpha}$ in the other ones. Denoting by $N(A_1, A_2)$ the correct weighted interpolator, we now have

$$N_\alpha(A_1, A_2) = M_{1-\alpha}(A_1, A_2) \text{ when } A_1\backslash A_2 \neq \emptyset \qquad (12.6)$$
$$= M_\alpha(A_1, A_2) \text{when } A_2\backslash A_1 \neq \emptyset$$

Figure 12.5 depicts such corrected interpolators.

The physical equation of the phenomenon Physically speaking, the accretion/ erosion process evolves at each instant from the stage it has reached before. It takes some $M_\alpha(X, Y)$, with $\alpha \in [0,1]$, as starting point and moves to $M_\beta[M_\alpha (X, Y), Y]$, for some value $\beta \in [0, 1]$. The weighted medians M_α do model this evolution because they form a semi-group. By calculating firstly the set $M_\alpha(X, Y)$ median between X and Y, and then the set $M_\beta[M_\alpha(X, Y), Y]$ between $M_\alpha(X, Y)$ and Y, we obtain indeed the same result as by calculating directly $M_\gamma(X, Y)$ for the weight $\gamma = \alpha + (1 - \alpha)\beta = \alpha + \beta - \alpha\beta$, i.e.

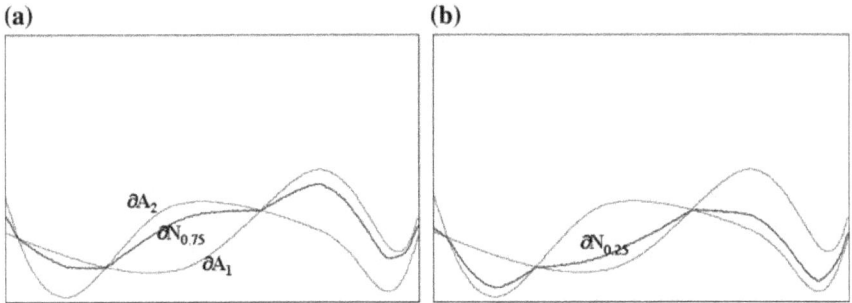

(a) **(b)**

Fig. 12.4 Raw weighted median lines

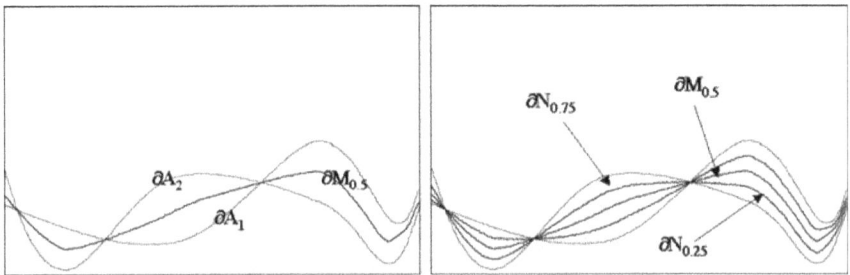

Fig. 12.5 Left: two shores A_1 and A_2, of boundaries ∂A_1 and ∂A_2, and their median line of boundary $\partial M_{0.5}$; right: the same, plus two additional weighted median lines according to Eq. (12.5)

$$M_\beta[M_\alpha(X, Y), Y] = M_{\alpha+\beta-\alpha\beta}(X, Y) \tag{12.7}$$

For example, in Fig. 12.5 right, the three median sets correspond to $\alpha = 0.75, 0.5$, and 0.25, and the weighted median $M_{0.75}$ is *also* the median element between $M_{0.5}$ and $A_1 \cup A_2$.

Proposition 3 *Given X, $Y \in \mathscr{K}'(R^n)$, the family $\{M_\alpha(X, Y), 0 \le \alpha \le 1\}$ of median elements form an additive semi-group for the addition $\alpha \otimes \beta = \alpha + \beta - \alpha\beta$.*

Proof Clearly, $\alpha \otimes \beta \in [0, 1]$, *thus Eq. (12.7) defines a commutative semi-group. The operation $\alpha \otimes \beta$ is also associative, since*

$$\gamma \otimes (\alpha + \beta - \alpha\beta) = \gamma + \alpha + \beta - \alpha\beta - \gamma\alpha - \gamma\beta + \gamma\alpha\beta$$

is symmetrical in α, β, γ, therefore $\alpha \otimes \beta$ is an algebraic addition.

Quench function and quench stripe As a matter of fact, the median operator provides *two outputs*, since we have on the one hand the (weighted or not) *median set M*, whose contour ∂M is the dark middle line in Fig. 12.5 left, or Fig. 12.6 left, and the *quench function q*, defined on ∂M and which gives at each the radius of the minimum disc hitting the two contours ∂A_1 and ∂A_2.

$$q(z) = \inf\{r : B_z(r) \cap \partial A_1 \ne \emptyset \text{ and } B_z(r) \cap \partial A_2 \ne \emptyset\} \tag{12.8}$$

A few of such discs, for the two inputs A_1 and A_2 of Fig. 12.3 left, are depicted in Fig. 12.6 left, and their union for the whole quench function gives the *quench stripe w*, i.e. the dark grey stripe W around the black line ∂M in Fig. 12.6 right, with

$$W = \cup\{B_z(q(z)), z \in M(A_1, A_2)\} \tag{12.9}$$

Note hat this dark grey stripe does not reach the edges of input sets A_1 and A_2, but an open version of their union, and a closed version of their intersection.

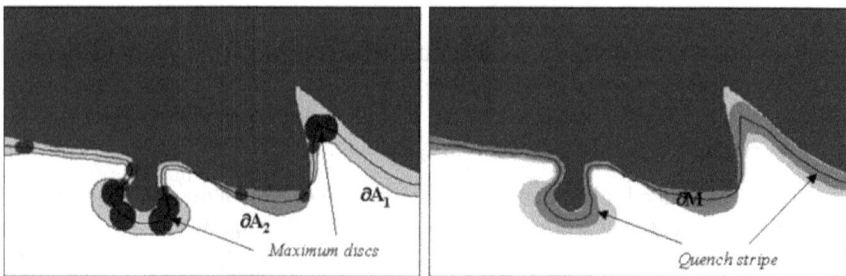

Fig. 12.6 Left: a few maximum discs centered on the median line; right: the dark grey stripe is the union of all maximum discs, or "quench stripe"

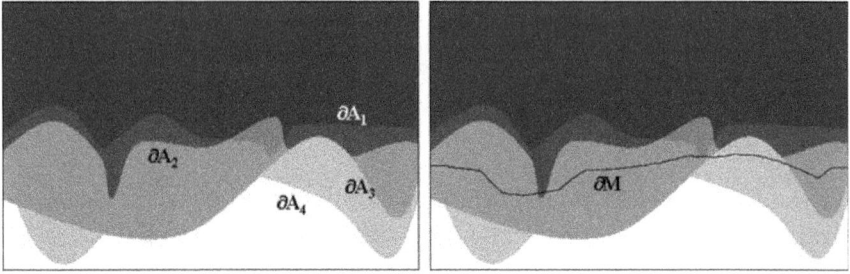

Fig. 12.7 Left: four shores; right in dark, their median line

Averages The structure of Eq. (12.7) suggests a technique for extending the median element to more than two input sets. Starting for example from the triplet $\{A_1, A_2, A_3\}$, we can calculate $M_{0.5}(A_1, A_2)$ in a first stage, and then $M_{0.33}[M_{0.5}(A_1, A_2), A_3]$. The resulting median element averages the three inputs, in a median sense. Figure 12.7 depicts an example of such an average for the four inputs $\{A_1, ..A_4\}$ shown in Fig. 12.7 left (two of them are the sets involved in Fig. 12.5 left). The initial stage consists in calculating $M_{0.5}(A_1, A_2)$ and $M_{0.5}(A_3, A_4)$, and the final one in calculating $M_{0.5}[M_{0.5}(A_1, A_2), M_{0.5}(A_3, A_4)]$, a set whose contour is drawn in black in Fig. 12.7 right. This final result is independent of the choice of the sets in the initial stage, and we could start as well from $M_{0.5}(A_1, A_3)$ and $M_{0.5}(A_2, A_4)$.

The averages obtained this way blur the structural features of the shores. Imagine for example that A_2, A_n are shifted versions of A_1 in the horizontal direction. As n increases, the median average contour tends towards an horizontal line: all features, gulfs, capes, etc. are lost. We meet here the same trouble as in interpolating moving objects, with translation and rotation. In case of shore movements, the translations are probably less intense, but the problem still remains. Remark also that this drawback is the counterpart of the advantage of preserving accretion and erosion zones.

12.4 Extrapolations via the Quench Function

In this section and the next one, we focus on the extrapolation of two shores at most, A_1 and A_2 say. If we dispose of a chronological sequence of the coast movements, A_1 and A_2 stand for the last two observations, A_2 being the more recent. The principle of the extrapolation consists in two possible changes:

1. that of the quench function according to a given law, which models the dynamics of the movement, and which results in a new quench stripe W;
2. that of the respective importances of A_1 and A_2. If we take the median $M_{0.5}(A_1, A_2)$, then both shores are given the same weight, but if we consider that A_2, more recent, is two times more significant than A_1, then we can take $N_{0.66}(A_1, A_2)$.

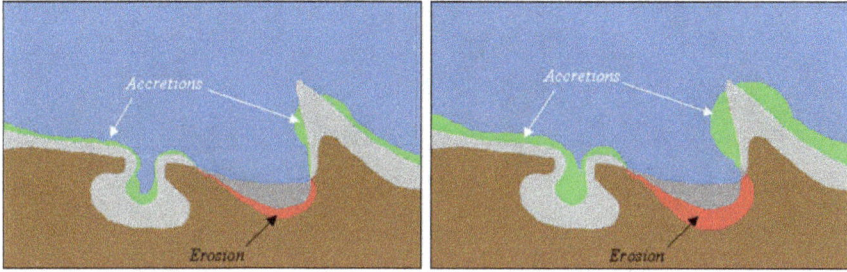

Fig. 12.8 Two extrapolations of the shoreline of Fig. 12.3; both are centered on $\partial M_{0.5}(A_1, A_2)$; the quench function is multiplied by 2 in the left image and by 3 in the right one

Fig. 12.8 depicts two extrapolations where the median element equals $M_{0.5}(A_1, A_2)$, hence where the two input shores are given the same importance, but where the quench stripe W of Eq. (12.9) is replaced by

$$W = \cup\{B_z(kq(z)),\ z \in M_{0.5}(A_1, A_2)\}$$

The radius of the disc centered at each point of $M_{0.5}(A_1, A_2)$ is quench value multiplied by factor k, with $k = 2$ for Fig. 12.8 left and $k = 3$ for Fig. 12.8 right. We see that, as k increases, both accretion and erosion zones are developed. We can also notice that the shape of the cape provokes a bizarre inflation in Fig. 12.8 right.

This swelling may be due to the great distance from the median line to extremity of the cape, as shown in Fig. 12.6 right, so that we can try to avoid it by making the median line closer to contour ∂A_2 which delineates the cape. Replace then the median set $M_{0.5}(A_1, A_2)$ by $N_\alpha(A_1, A_2)$, in the sense of Eq. (12.6), with $\alpha = 0.75$, so that the quench stripe becomes

$$W = \cup\{B_z(kq(z)),\ z \in N_{0.75}(A_1, A_2)\}.$$

The resulting changes are depicted in Fig. 12.9, left for $k = 3$, and right for $k = 4$. By comparing Figs. 12.8 right and 12.9 left where the quench function is multiplied by the same value $k = 3$, we see that the cape inflates less, but in compensation the erosion zone vanished. The erosion can reappear by taking $k = 4$ (Fig. 12.9 right), but again the cape inflates as strongly as in the previous extrapolation of Fig. 12.8 right.

In fact, transforming a quench function according to pure magnification is probably too poor. One can easily imagine more sophisticated laws such as the two following ones:

1. the median line is slightly moved toward the second contour, by taking $N_{0.66}$ (A_1, A_2), and the quench stripe W is obtained by dilating each point z of the median line by the disc of radius $2q(z)$ and by the segment $L_\alpha(2q(z))$ of length $2q(z)$ in the main direction α of the cape, which gives

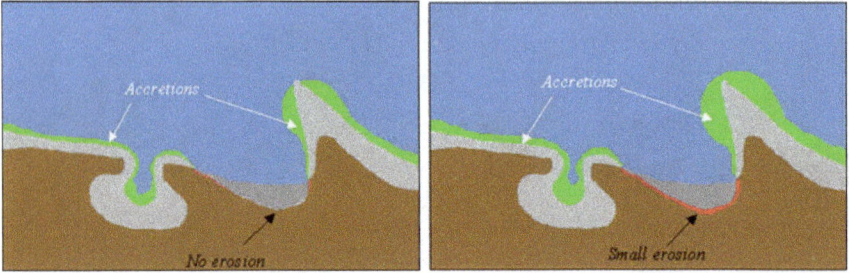

Fig. 12.9 Two other extrapolations of the shoreline of Fig. 12.3; both are centered on $\partial N_{0.75}(A_1, A_2)$; the quench function is multiplied by 3 in the left image and by 4 in the right one

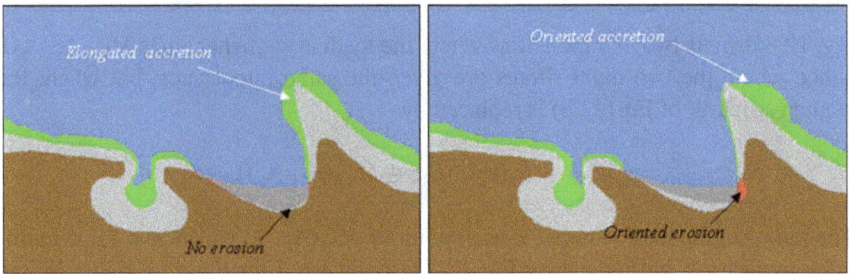

Fig. 12.10 Two extrapolations of the shoreline of Fig. 12.3, by emphazising the new capes in the left image, and by introducing an east-west trend in the right one

$$W = \cup\{[B_z(2q(z) \oplus L_\alpha(2q(z))], \; z \in N_{0.66}(A_1, A_2)\}$$

and which is depicted in Fig. 12.10 left. The accretion around the cape turns out to be now more realistic, but the erosion zone has disappeared.

2. The median set $N_{0.66}(A_1, A_2)$ is left unchanged, and a supplementary trend in the horizontal direction is introduced by a dilating points z by the horizontal segment $L_0(3q(z))$. For avoiding too fast changes, the parameters of the two other dilations are divided by 2. The shifting effect of the trend operation appears clearly in Fig. 12.10 right, where the accretion forms a deposit at the east of the cape. Similarly, the directional effect of the erosion holds for west oriented regions.

Unlike the previous models, which all are invariant under rotation of the map, these last two laws, which model marine currents, depend on the North direction (see Fig. 12.10).

12.5   Accretion and Homotopy

It may happen that, for some reasons, one wishes to preserve the homotopy of the shore, which excludes the creation, or the suppression, of lakes and islands. Now, by dilating enough the shore of Fig. 12.10, we risk to close the gulf on the left and to generate in internal island. An easy way to protect the gulf as such consists in replacing the dilation w.r.t. the unit disc by a cycle of elementary homotopic thickenings in the eight directions of the square grid, or the six ones in the hexagonal case (Serra 1982). The circular dilation of size n becomes the series of n thickening cycles. One can see in Figure ll, left and right, the results of two thickenings of sizes 25 and 33 respectively (for a 512×320 digital image). The gulf is preserved by a narrow channel, which could be enlarged by modifying the homotopy preservation algorithm. This conceptually simple method is not the only possible one. In Vidal et al. (2005) the authors propose a median set based interpolation that preserves particles by marking them by a homotopic thinning, and translating them during the interpolation process.

12.6   Conclusion

Our purpose was to demonstrate the physical sense of the median set approach and its flexibility. In the first section, we indicated three features to be respected by interpolations. According to the first one, an accretion (resp. erosion) zone must continue to evolve by accretion (resp. erosion). This basic modality is fulfilled by all models of Sect. 12.4. The laws proposed in this section are far from being the only possible ones. In particular, each of the six examples of the section is given a same law for accretion and erosion, which is not at all an obligation. The second feature holds for the role of the past. In the approach of Sect. 12.4, this past reduces to the last two stages: they suffice to determine the starting shoreline, the "gradient", and the location of accretion/erosion (Fig. 12.11).

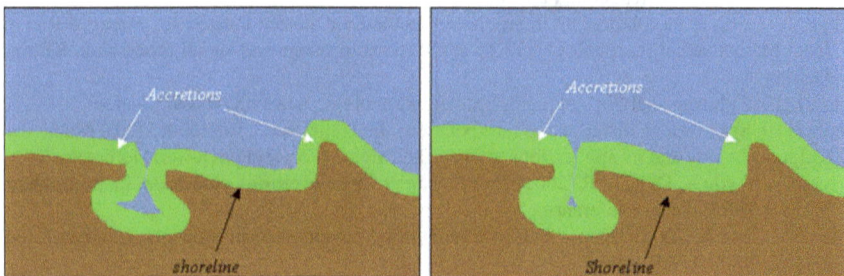

Fig. 12.11 Two extrapolations of the shoreline of Fig. 12.3 by homotopic thickenings of sizes 25 (left) and 33 (right)

The third feature was the subject of Sect. 12.5, where a thickening is substituted for the dilation in the extrapolator, in order to preserve homotopy. Indeed, all extrapolation equations, from sections two to four, can be rewritten by replacing the unit disc erosion and dilation by unit cycles of thinnings and thickenings, and the linear dilations by unidirectional thickenings. It would result in a series of algorithms where increasingness is lost (non direct extension to numerical functions) but where topological features are preserved.

Finally, as the weighted median of Eq. (12.4) is an increasing function of its two operands, it extends to numerical functions by means of their subgraphs, and allows to process colour images (Daya Sagar 2007).

Acknowledgements I am extremely grateful to Dr B.R. Kiran for his precious help in preparing this chapter.

References

Angulo J, Meyer F (2009) Morphological exploration of shape spaces. In: Wilkinson MHF, Roerdink JBTM (eds) Mathematical morphology and its applications to signal and image processing. Springer, Berlin, pp 226–237

Beucher S (1998) Interpolation of sets, of partitions and of functions. In: Heijmans H, Roerdink J (eds) Mathematical morphology and its applications to image and signal processing. Kluwer, Dordrecht

Casas JR (1996) Image compression based on perceptual coding techniques, PhD thesis, UPC Barcelona, March 1996

Charpiat G, Faugeras O, Keriven R, Maurel P (2006) Distance-based shape statistics. IEEE ICASP 5:925–928

Chaussard J (2006) Surveillance cotire l'aide de la morphologie, a bibliographical study, A2SI, ESIEE France

Daya Sagar BS (2007) Universal scaling laws in surface water bodies and their zones of influence. Water Resour Res 43(2):W02416

Huttenlocher DP, Klunderman GA, Rucklidge WJ (1995) Comparing images using the Hausdorff distance. IEEE PAMI. 15(9)

Iwanowski M, Serra J (2000) The morphological-affine object deformation. In: Goutsias J, Vincent L, Bloomberg DS (eds) Mathematical morphology and its applications to image and signal processing. Kluwer Academic Publisher, Dordrecht, pp 81–90

Meyer F (1996) A morphological interpolation method for mosaic images. In: Maragos P et al. (eds) Mathematical morphology and its applications to image and signal processing. Kluwer, Dordrecht

Serra J (1982) Image analysis and mathematical morphology. Academic Press, London

Serra J (1998) Hausdorff distance and Interpolations. In: Heijmans H, Roerdink J (eds) Mathematical morphology and its applications to image and signal processing. Kluwer, Dordrecht

Srivastava A, Niu X, Di K, Li R (2005) Shoreline modeling and erosion prediction. In: proceedings of the ASPRS annual conference

Thao NV, Chen X (2005) Temporal GIS for monitoring lagoon inlets movements. In: Chen X (ed) DMAI' 2005. AIT Bangkok

Vidal J, Crespo J, Maojo V (2005) Recursive interpolation technique for binary images based on morphological median sets. In: Ronse Ch, Najman L, Decencire E (eds) Mathematical morphology: 40 years on. Springer, Berlin, pp 53–65

13

Satellite Image Analysis and Geostatistical Tools

A. F. Militino, M. D. Ugarte and U. Pérez-Goya

Abstract Satellite remote sensing data have become available in meteorology, a-griculture, forestry, geology, regional planning, hydrology or natural environment sciences since several decades ago, because satellites provide routinely high quality images with different temporal and spatial resolutions. Joining, combining or smoothing these images for a better quality of information is a challenge not always properly solved. In this regard, geostatistics, as the spatio-temporal stochastic techniques of geo-referenced data, is a very helpful and powerful tool not enough explored in this area yet. Here, we analyze the current use of some of the geostatistical tools in satellite image analysis, and provide an introduction to this subject for potential researchers.

13.1   Introduction

The spatio-temporal analysis of satellite remote sensing data using geostatistical tools is still scarce when comparing with other kinds of analyses. In this chapter we provide an introduction to this field for geostatisticians, empathising the importance of using the spatio-temporal stochastic methods in satellite imagery and providing

A. F. Militino (✉) · M. D. Ugarte · U. Pérez-Goya
Department of Statistics and O.R., Public University of Navarra (Spain),
Pamplona, Spain
e-mail: militino@unavarra.es

U. Pérez-Goya
e-mail: unai.perez@unavarra.es

A. F. Militino · M. D. Ugarte
InaMat (Institute for Advanced Materials), Pamplona, Spain
e-mail: lola@unavarra.es

a review of some applications (Sagar and Serra 2010). We explain how to proceed for accessing remote sensing data, and which are the common tools for download-ing, pre-processing, analysing, interpolating, smoothing and modeling these data. The chapter encloses six additional sections where a short explanation of the state of the art in the analysis of remote sensing data using free statistical software is giv-en. Particular attention is devoted to the use of geostatistical tools in this subject. Section 13.2 explains the profile and the main features of the most popular satel-lites. It also encompasses Sect. 13.2.1 for describing some R packages for importing, analysing, and managing satellite images. Section 13.3 explains how to retrieve two derived variables, the normalized difference vegetation index (NDVI) and the land surface temperature (LST). In Sect. 13.4 some common methods of pre-processing data after downloading satellite images are reviewed. Section 13.5 explains the im-portance of the spatial interpolation in remote sensing data and reviews the most pop-ular interpolation methods. The actual scenario of the spatio-temporal geostatistics is reviewed in Sect. 13.6, where an additional subsection describes some *R* packages for using spatial and spatio-temporal geostatistics techniques with satellite images. The paper ends up with some conclusions in Sect. 13.7.

13.2   Satellite Images

Satellite images are available since more than four decades ago, and since then there has been a notable improvement in quality, quantity, and accessibility of these im-ages, making it easier to extract huge amounts of data from all over the Earth. We can retrieve data from the land or the ocean, from the coast or the mountains, and also from the atmosphere where advanced sensors give the opportunity of monitor-ing meteorological variables that are crucial for the study of the climatic change, the phenology trend, the changes in vegetation or many other environmental processes.

Remote sensing refers to the process of acquiring information from the Earth or the atmosphere using sensors or space shuttles platforms. Therefore, remote sensing is born as a crucial necessity when using satellite images for analyzing and convert-ing them into different frames of data that can be managed with specific software. Nowadays, Landsat, Modis, Sentinel or Noaa are some of the most popular satellite missions among researchers and practitioners of remote sensing data because of the free accessibility. Next, we summarize the main characteristics of these missions:

1. LANDSAT, meaning Land+Satellite, represents the world's longest continuous-ly acquired collection of space-based moderate-resolution land remote sensing data. See GLCF (2017) for details. It is available since 1972 from six satel-lites in the Landsat series. These satellites have been a major component of NASA's Earth observation program, with three primary sensors evolving over thirty years: MSS (Multi-spectral Scanner), TM (Thematic Mapper), and ETM+

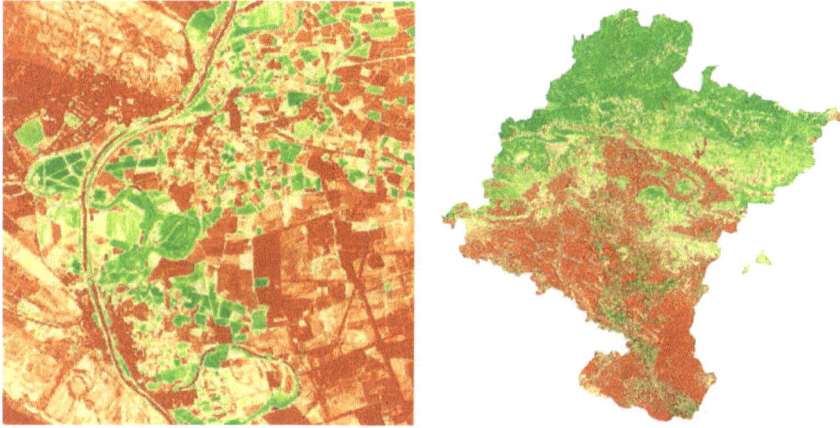

Fig. 13.1   (Left) NDVI Sentinel image of Funes village in Navarra, and (Right) NDVI for the whole Navarra (Spain)

which is closely related to the presence of vegetation. Although numerical limits of NDVI can vary for the vegetation classification, it is widely accepted that negative NDVI values correspond to water or snow. NDVI values close to zero could correspond to bare soils, yet these soils can show a high variability. Values between 0.2 and 0.5 (approximately) to sparse vegetation, and values between 0.6 and 1.0 conform to dense vegetation such as that found in temperate and tropical forests or crops at their peak growth stage. Therefore, NDVI provides a very valuable instrument for monitoring crops, vegetation, and forestry, and it is directly calculated in specific images by the aforementioned satellites missions. On the left of Fig. 13.1 a Sentinel NDVI satellite image of Funes, a village of Navarra (Spain) is shown, and on the right of the same Figure, the NDVI for the whole region of Navarra.

Another important variable derived with satellite images is the land surface temperature (LST), that can be retrieved with different algorithmic procedures. As an example Sobrino et al. (2004) compare three methods to retrieve the LST from thermal infrared data supplied by band 6 of the Thematic Mapper (TM) sensor onboard the Landsat 5 satellite. The first is based on the radiative transfer equation using in situ radiosounding data. The others are the mono-window algorithm developed by Qin et al. (2001) and the single-channel algorithm developed by Jiménez-Muñoz and Sobrino (2003). Many satellites platforms provide specific images of LST all over the Earth, because it is also a very outstanding variable for many environmental process. Figure 13.2 shows the daily land surface temperature in Navarra (Spain) the 13th of July 2015 from TERRA satellite.

Fig. 13.2   Land Surface Temperature of Navarra the 13th of July 2015

13.4   Pre-processing

The atmosphere is between the satellite and the Earth, and its effects over the electromagnetic radiation caused by the satellite can distort, blur or degrade the images. These effects must be corrected before the image processing. The correction consists of composing several images into a new single one. Different algorithms have been developed in the literature according to the derived variable. The most common method with NDVI is the maximum value composite (MVC) procedure (Holben 1986) that assigns the maximum value of the time-series of pixels across the composite period. Alternative techniques include using a bidirectional reflectance distribution function (BRDF-C) to select observations and the constraint view angle maximum value composite (CV-MVC) (MODIS 2017). For LST day/night it is common to average the cloud-free pixels over the compositing period (Vancutsem et al. 2010). Nowadays, many composite images can be directly downloaded with different spatial and temporal resolutions. For example, raw daily images can be downloaded from AQUA or TERRA satellites all over the world, but usually composite images are at least of weekly or bi-weekly temporal resolution.

Spatial and temporal resolutions are also different from the same or different satellites. High temporal resolution can be useful when tracking seasonal changes in vegetation on continental and global scales, but when downscaling to small regions, a higher spatial resolution is needed, and frequently with lower temporal resolution. At this step, numerical, physical or mechanical analyses solve the image pre-processing. Later, removing the effect of clouds or other atmospheric effects is also required, otherwise remote sensing data can be inaccurate. Sometimes, the highest presence of clouds determine the dropout of several images, but if they are only partially clouded, different approaches for eliminating these effects can be used. Noise reduction in image time series is neither simple nor straightforward. Many alternatives have been provided. For example R.HANTS macro of GRASS, SPIRITS, BISE, TIMESAT, GAPFILL or the CACAO methods are very well spread. R.HANTS performs an harmonic analysis of time series in order to estimate missing values and identify outliers (Roerink et al. 2000). SPIRITS is a software that processes time series of images (Eerens et al. 2014). It was developed by PROBA-V data provider and gives four smoothing options, including MEAN (Interpolate missing values & apply Running Mean Filter RMF) and BISE (Best Index Slope Extraction), (Viovy et al. 1992). TIMESAT uses numerical procedures based on Fourier analysis, Gauss, double logistic or SavitzkyGolay filters (Jönsson and Eklundh 2004). GAPFILL uses quantile regression to produce smoothed images where the effect of the clouds have been reduced. Usually, every software has different requirements with regard to the number of images necessary for smoothing (Atkinson et al. 2012). Finally, CACAO software (Verger et al. 2013) provides smoothing, gap filling, and characterizing seasonal anomalies in satellite time series.

All these procedures give composite images that are smoothed versions of the raw images, but very often they are not completely free of noise. Many of the attributes that can be extracted from the combination of satellite image bands are still vulnerable to many atmospheric or electronic accidents. For example, highly reflective surfaces, including snow and clouds, and sun-glint over water bodies may saturate the reflective wavelength bands, with saturation varying spectrally and with the illumination geometry (Roy et al. 2016). Land surface temperature or normalized vegetation index are examples of attributes where these type of errors can be present. Therefore, after pre-processing is done, interpolation and smoothing methods can be very useful for drawing or detecting trend changes, clustering or many other processes on remote sensing data.

13.5   Spatial Interpolation

Likely, interpolation and classification are among the most used tools with remote sensing data. Classification of satellite images in supervised or unsupervised versions are important research areas not only with satellite images but also with big data and data mining where there are a great number of algorithmic procedures (Benz et al. 2004). Here, we are more interested in interpolation as it is more closely related to geostatistics.

Interpolation has been widely used in environmental sciences. Li and Heap (2011) revise more than 50 different spatial interpolation methods that can be summarized in three categories: non-geostatistical methods, geostatistical methods, and combined methods. All of them can be represented as weighted averages of sampled data. Among the non-geostatistical methods the authors find: nearest neighbours, inverse distance weighting, regression models, trend surface analysis, splines and local trend surfaces, thin plate splines, classification, and regression trees. The different versions of simple, ordinary, disjunctive or model-based kriging are among the geostatistical methods. The combined methods include: trend surface analysis combined with kriging, linear mixed models, regression trees combined with kriging or regression kriging.

Recently, Jin and Heap (2014) present an excellent review of spatial interpolation methods in environmental sciences introducing 10 methods from the machine learning field. These methods include support vector machines (SVM), random forests (RF), neural networks, neuro-fuzzy networks, boosted decision trees (BDT), the combination of SVM with inverse distance weighting (IDW) or ordinary kriging (OK), the combination of RF with IDW or OK (RFIDW, RFOK), general regression neural network (GRNN), the combination of GRNN with IDW or OK, and the combination of BDT with IDW or OK. Although all these methods were not developed specifically for remote sensing data, nowadays the majority of them have been implemented in different packages of the free statistical software R, and can be used with satellite images. Many of these methods are ready to use and interpret, but the family of kriging methods as the core of geostatistics, are preferred and widely used.

13.6   Spatio-Temporal Interpolation

Since the publication of the seminal book *Spatial Autocorrelation* (Cliff and Ord 1973), and at latter date *Spatial Statistics* (Ripley 1981), *Statistics for Spatial Data* (Cressie and Wikle 2015), and *Multivarate Geostatistics* (Wackernagel 1995) books, there has been a rapid growth of spatial geostatistical methods, as they are essential tools for interpolating meteorological, physical, agricultural or environmental variables in locations where these variables are not observed.

The use of spatial geostatistics with remote sensing data is also very well widespread, and its procedures are present in many specific softwares of satellite image analysis (Stein et al. 1999). Geostatistics techniques can help to explore and describe the spatial variability, to design optimum sampling schemes, and to increase the accuracy estimation of the variables of interest. These models can be enriched with auxiliary information coming from classified land cover or historical information (Curran and Atkinson 1998). Kriging is the most popular geostatistical method with several versions such as block kriging, universal kriging, ordinary kriging, regression kriging or indicator kriging. It provides the spatial interpolation of different spatial variables through the use of spatial stochastic models, and it is the best linear unbiased predictor under normality assumptions when using spatially dependent data.

However, the extension to the spatio-temporal geostatistics methods is more complicated. Time series models typically assume a regularly sampling over time, but the temporal lag operator cannot be easily generalized to the spatial domain, where data are likely irregularly sampled (Phaedon and André 1999). Scales of time and space are different, therefore defining joint spatio-temporal covariance functions is not a trivial task (De Iaco et al. 2002). Recently, Cressie and Wikle (2015) show the state of the art in this area and explain the difficulties of inverting covariance matrices in spatio-temporal kriging, because it becomes problematic without some form of separable models or dimension reduction. Modelling the spatio-temporal dependence is frequently case-specific. Therefore, yet the presence of the spatio-temporal keyword is abundant in many satellite imagery papers, the use of spatio-temporal stochastic models is scarce. Very often, spate-time refers only to descriptive analyses of time series of satellite images where every image is analyzed as a set of separate pixels, i.e., when estimating trends, or trend changes, statistical methods of univariate time series are used for every pixel. For example, when completing, reconstructing or predicting the spatial and temporal dynamics of the future NDVI distribution many papers use a time series of images (Forkel et al. 2013; Tüshaus et al. 2014; Klisch and Atzberger 2016; Wang et al. 2016; Liu et al. 2015; Maselli et al. 2014). These studies include temporal correlation of individual pixels at different resolutions but ignoring spatial dependence among them.

Spatio-temporal stochastic models use the spatial or temporal dependence to estimate optimally local values from sampled data. In satellite images, sampled data can be a huge amount of spatially and temporally dependent pixels, if a sequence of images is involved. We briefly review in what follows some stochastic spatio-temporal models that can be used when analysing remote sensing data.

1. Spatio-temporal kriging (Gasch et al. 2015). This paper uses spatio-temporal R packages for fitting some of the following spatio-temporal covariance functions: separable, product-sum, metric and sum-metric classesin a spatio-temporal kriging model, and a random forest algorithm for modeling dynamic soil properties in 3-dimensions.
2. State-space models (Cameletti et al. 2011). The authors apply a family of state-space models with different hierarchical structure and different spatio-temporal covariance function for modelling particular matter in Piemonte (Italy).
3. Hierarchical spatio-temporal model (Cameletti et al. 2013). The paper introduces a hierarchical spatio-temporal model for particulate matter (PM) concentration in the North-Italian region Piemonte. The authors use stat-space models involving a Gaussian Field (GF), affected by a measurement error, and a state process characterized by a first order autoregressive dynamic model and spatially correlated innovations. The estimation is based on Bayesian methods and

consists of representing a GF with Matérn covariance function as a Gaussian Markov Random Field (GMRF) through the Stochastic Partial Differential E- quations (SPDE) approach. Then, the Integrated Nested Laplace Approximation (INLA) algorithm is proposed as an alternative to MCMC methods, giving rise to additional computational advantages (Rue et al. 2009).

4. Spatio-temporal data-fusion (STDF) methodology (Nguyen et al 2014). This method is based on reduced-dimensional Kalman smoothing. The STDF is able to combine the complementary GOSAT and AIRS datasets to optimally estimate lower-atmospheric CO_2 mole fraction over the whole globe.

5. Hierarchical statistical model (Kang et al. 2010). This model includes a spatio-temporal random effects (STRE) model as a dynamical component, and a tem- porally independent spatial component for the fine-scale variation. This article demonstrates that spatio-temporal statistical models can be made operational and provide a way to estimate level-3 values over the whole grid and attach to each value a measure of its uncertainty. Specifically, a hierarchical statistical model is presented, including a spatio-temporal random effects (STRE) mod- el as a dynamical component and a temporally independent spatial component for the fine-scale variation. Optimal spatio-temporal predictions and their mean squared prediction errors are derived in terms of a fixed-dimensional Kalman filter.

6. Three-stage spatio-temporal hierarchical model (Fassò and Cameletti 2009). This work gives a three-stage spatio-temporal hierarchical model including spatio-temporal covariates. It is estimated through an EM algorithm and boot- strap techniques. This approach has been used by (Militino et al. 2015) for in- terpolating daily rainfall data, and for estimating spatio-temporal trend changes in NDVI with satellite images of Spain from 2011-2013 (Militino et al. 2017).

7. Space-varying regression model (Bolin et al. 2009). In this space-varying regres- sion model the regression coefficients for the spatial locations are dependent. A second order intrinsic Gaussian Markov Random Field prior is used to specify the spatial covariance structure. Model parameters are estimated using the Ex- pectation Maximisation (EM) algorithm, which allows for feasible computation times for relatively large data sets. Results are illustrated with simulated data sets and real vegetation data from the Sahel area in northern Africa.

13.6.1 Geostatistical R Packages

In this section we briefly describe some of the most useful R packages for geostatisti- cal analysis, including spatial and spatio-temporal interpolation in satellite imagery.

1. FRK (Cressie and Johannesson 2008) means fixed rank kriging and it is a tool for spatial/spatio-temporal modelling and prediction with large datasets.

2. geoR (Ribeiro Jr et al. 2001) offers classical geostatistics techniques for analysing spatial data. The extension to generalized linear models was made in geoRglm package (Christensen and Ribeiro 2002).

3. georob (R Core Team 2017) fits linear models with spatially correlated errors to geostatistical data that are possibly contaminated by outliers.

4. geospt (Melo et al. 2012) estimates the variogram through trimmed mean and does summary statistics from cross-validation, pocket plot, and design of optimal sampling networks through sequential and simultaneous points methods.

5. geostatsp (Brown 2015) provides geostatistical modelling facilities using raster. Non-Gaussian models are fitted using INLA, and Gaussian geostatistical models use maximum likelihood estimation.

6. gstat (Pebesma 2004) does spatio-temporal kriging, sequential Gaussian or indicator (co)simulation, variogram and variogram map plotting utility functions.

7. RandomFields (Schlather et al. 2015) provides methods for the inference on and the simulation of Gaussian fields.

8. spacetime (Pebesma et al. 2012) gives methods for representations of spatio-temporal sensor data, and results from predicting (spatial and/or temporal interpolation or smoothing), aggregating, or sub-setting them, and to represent trajectories.

9. spatial (Venables and Ripley 2002) provides functions for kriging and point pattern analysis.

10. spatialEco (Evans 2016) does spatial smoothing, multivariate separability, point process model for creating pseudo- absences and sub-sampling, polygon and point-distance landscape metrics, auto-logistic model, sampling models, cluster optimization and statistical exploratory tools. It works with raster data.

11. SpatialTools (R Core Team 2017) contains tools for spatial data analysis with emphasis on kriging. It provides functions for prediction and simulation.

12. spBayes (Finley et al. 2007) fits univariate and multivariate spatio-temporal random effects models for point-referenced data using Markov chain Monte Carlo (MCMC).

13.7 Conclusions

The multitemporal Earth observation satellites have been very well developed since the seventies, and along with the free availability of millions of satellite images, the number of publications of remote sensing data with geostatistical techniques has been rapidly increased. But unfortunately, not all published papers deriving, analysing or monitoring spatio-temporal evolutions, spatio-temporal trends or spatio-temporal changes are necessarily geostatistical papers, because they do not really use spatio-temporal stochastic models. These models are still scarce in remote sensing data because many of these models are computationally very intensive, or because they are not so broadly applicable as the spatial models are. The solutions found in the literature are very well fitted to specific problems, but we cannot always plug-in to other applications. The use of time series analysis in remote sensing opens a great window of opportunities for monitoring, smoothing, and detecting changes in large series of satellite images, but there are still many remote sensing papers ignoring the spatial dependence when analysing time series of images (Ban 2016). Instead, a huge discretization of the problem is presented where time-series of pixels are treated as spatially independent.

Nowadays, the upcoming opportunities for geostatisticians in remote sensing data are not based on the use of spatial models and time series separately, but on the use of spatial, temporal, or spatio-temporal stochastic models embedding both types of dependencies when necessary. Moreover, a single free statistical software like R is a powerful tool for downloading, importing, accessing, exploring, analysing and running advanced statistical modelling with remote sensing data in a row.

Acknowledgements This research was supported by the Spanish Ministry of Economy, Industry and Competitiveness (Project MTM2017-82553-R), the Government of Navarra (Project PI015, 2016 and Project PI043 2017), and by the Fundación Caja Navarra-UNED Pamplona (2016 and 2017).

References

Aschbacher J, Milagro-Pérez MP (2012) The European earth monitoring (GMES) programme: status and perspectives. Remote Sens Environ 120:3–8

Atkinson PM, Jeganathan C, Dash J, Atzberger C (2012) Inter-comparison of four models for smoothing satellite sensor time-series data to estimate vegetation phenology. Remote Sens Environ 123:400–417

Ban Y (2016) Multitemporal remote sensing. Methods and applications, vol 1. Remote sensing and digital image processing. Springer, Berlin

Benz UC, Hofmann P, Willhauck G, Lingenfelder I, Heynen M (2004) Multi-resolution, object-oriented fuzzy analysis of remote sensing data for GIS-ready information. ISPRS J Photogramm Remote Sens 58(3):239–258

Bolin D, Lindström J, Eklundh L, Lindgren F (2009) Fast estimation of spatially dependent temporal vegetation trends using Gaussian Markov random fields. Comput Stat Data Anal 53(8):2885–2896

Brown PE (2015) Model-based geostatistics the easy way. J Stat Softw 63(12):1–24. http://www.jstatsoft.org/v63/i12/

Cameletti M, Ignaccolo R, Bande S (2011) Comparing spatio-temporal models for particulate matter in Piemonte. Environmetrics 22(8):985–996

Cameletti M, Lindgren F, Simpson D, Rue H (2013) Spatio-temporal modeling of particulate matter concentration through the SPDE approach. AStA Adv Stat Anal 97(2):109–131

Christensen O, Ribeiro PJ (2002) geoRglm - a package for generalised linear spatial models. R-news 2(2):26–28. http://cran.R-project.org/doc/Rnews. ISSN 1609-3631

Cliff AD, Ord JK (1973) Spatial autocorrelation, vol 5. Pion, London

Cressie N, Johannesson G (2008) Fixed rank kriging for very large spatial data sets. J R Stat Soci: Ser B (Stat Methodol) 70(1):209–226

Cressie N, Wikle CK (2015) Statistics for spatio-temporal data. Wiley, New York

Curran PJ, Atkinson PM (1998) Geostatistics and remote sensing. Prog Phys Geogr 22(1):61–78

De Iaco S, Myers DE, Posa D (2002) Nonseparable space-time covariance models: some parametric families. Math Geol 34(1):23–42

Eerens H, Haesen D, Rembold F, Urbano F, Tote C, Bydekerke L (2014) Image time series processing for agriculture monitoring. Environ Model Softw 53:154–162

Evans JS (2016) spatialEco. http://CRAN.R-project.org/package=spatialEco, R package version 0.0.1-4

Fassò A, Cameletti M (2009) The EM algorithm in a distributed computing environment for modelling environmental space-time data. Environ Model Softw 24(9):1027–1035

Finley AO, Banerjee S, Carlin BP (2007) spBayes: an R package for univariate and multivariate hierarchical point-referenced spatial models. J Stat Softw 19(4):1–24. http://www.jstatsoft.org/v19/i04/

Forkel M, Carvalhais N, Verbesselt J, Mahecha MD, Neigh CS, Reichstein M (2013) Trend change detection in NDVI time series: effects of inter-annual variability and methodology. Remote Sens 5(5):2113–2144

Gasch CK, Hengl T, Gräler B, Meyer H, Magney TS, Brown DJ (2015) Spatio-temporal interpolation of soil water, temperature, and electrical conductivity in 3D + T: the cook agronomy farm data set. Spat Stat 14:70–90

Gerber F, Furrer R, Schaepman-Strub G, de Jong R, Schaepman ME (2016) Predicting missing values in spatio-temporal satellite data. arXiv:1605.01038

GLCF (2017) Global land cover facility. http://glcf.umd.edu/data/landsat/

Goslee SC (2011) Analyzing remote sensing data in R: the landsat package. J Stat Softw 43(4):1–25. http://www.jstatsoft.org/v43/i04/

Holben BN (1986) Characteristics of maximum-value composite images from temporal AVHRR data. Int J Remote Sens 7(11):1417–1434

Jiménez-Muñoz JC, Sobrino JA (2003) A generalized single-channel method for retrieving land surface temperature from remote sensing data. J Geophys Res: Atmos 108 (D22)

Jin L, Heap AD (2014) Spatial interpolation methods applied in the environmental sciences: a review. Environ Model Softw 53:173–189

Jönsson P, Eklundh L (2004) Timesat a program for analyzing time-series of satellite sensor data. Comput Geosci 30(8):833–845

Kang EL, Cressie N, Shi T (2010) Using temporal variability to improve spatial mapping with application to satellite data. Can J Stat 38(2):271–289

Klisch A, Atzberger C (2016) Operational drought monitoring in Kenya using modis NDVI time series. Remote Sens 8(4):267

Li J, Heap AD (2011) A review of comparative studies of spatial interpolation methods in environmental sciences: performance and impact factors. Ecol Inform 6(3):228–241

Liu Y, Li Y, Li S, Motesharrei S (2015) Spatial and temporal patterns of global NDVI trends: correlations with climate and human factors. Remote Sens 7(10):13,233–13,250

Maselli F, Papale D, Chiesi M, Matteucci G, Angeli L, Raschi A, Seufert G (2014) Operational monitoring of daily evapotranspiration by the combination of MODIS NDVI and ground meteorological data: application and evaluation in Central Italy. Remote Sens Environ 152:279–290

Matzke NJ (2013) modiscloud: an R Package for processing MODIS Level 2 Cloud Mask products. University of California, Berkeley, Berkeley, CA, http://cran.r-project.org/web/packages/modiscloud/index.html, this code was developed for the following paper: Goldsmith, Gregory; Matzke, Nicholas J.; Dawson, Todd (2013). The incidence and implications of clouds for cloud forest plant water relations. Ecol Lett, 16(3), 307–314. https://doi.org/10.1111/ele.12039

Maus V, Camara G, Cartaxo R, Sanchez A, Ramos FM, de Queiroz GR (2016) A time-weighted dynamic time warping method for land-use and land-cover mapping. IEEE J Sel Top Appl Earth Obs Remote Sens 9(8):3729–3739. https://doi.org/10.1109/JSTARS.2016.2517118

Melo C, Santacruz A, Melo O (2012) geospt: an R package for spatial statistics. http://geospt.r-forge.r-project.org/, R package version 1.0-0

Militino A, Ugarte M, Goicoa T, Genton M (2015) Interpolation of daily rainfall using spatiotemporal models and clustering. Int J Climatol 35(7):1453–1464

Militino AF, Ugarte MD, Pérez-Goya U (2017) Stochastic spatio-temporal models for analysing NDVI distribution of GIMMS NDVI3g images. Remote Sens 9(1):76

MODIS (2017) https://modis.gsfc.nasa.gov/about/

Nauss T, Meyer H, Detsch F, Appelhans T (2015) Manipulating satellite data with satellite. www.environmentalinformatics-marburg.de

Nguyen H, Katzfuss M, Cressie N, Braverman A (2014) Spatio-temporal data fusion for very large remote sensing datasets. Technometrics 56(2):174–185

NOAA (2017) National Oceanic and Atmospheric Administration. https://www.nesdis.noaa.gov/

Pebesma E et al (2012) Spacetime: spatio-temporal data in R. J Stat Softw 51(7):1–30

Pebesma EJ (2004) Multivariable geostatistics in S: the gstat package. Comput Geosci 30(7):683–691

Phaedon CK, André GJ (1999) Geostatistical spacetime models: a review. Math Geol 31(6):651–684

Qin Z, Karnieli A, Berliner P (2001) A mono-window algorithm for retrieving land surface temperature from landsat TM data and its application to the Israel-Egypt border region. Int J Remote Sens 22(18):3719–3746

R Core Team (2017) R: a language and environment for statistical computing. R Foundation for Statistical Computing, Vienna, Austria, https://www.R-project.org/

Ribeiro PJ Jr, Diggle PJ et al (2001) geoR: a package for geostatistical analysis. R news 1(2):14–18

Ripley BD (1981) Spatial statistics, vol 575. Wiley, New York

Roerink G, Menenti M, Verhoef W (2000) Reconstructing cloudfree NDVI composites using Fourier analysis of time series. Int J Remote Sens 21(9):1911–1917

Rouse J Jr, Haas R, Schell J, Deering D (1974) Monitoring vegetation systems in the Great Plains with ERTS. NASA special publication 351:309

Roy D, Kovalskyy V, Zhang H, Vermote E, Yan L, Kumar S, Egorov A (2016) Characterization of landsat-7 to landsat-8 reflective wavelength and normalized difference vegetation index continuity. Remote Sens Environ 185:57–70

Rue H, Martino S, Chopin N (2009) Approximate Bayesian inference for latent Gaussian models by using integrated nested laplace approximations. J R Stat Soc: Ser B (Stat Methodol) 71(2):319–392

Sagar DB, Je Serra (2010) Spacial issue on spatial information retrieval, analysis, reasoning and modelling. Int J Remote Sens 31(22):5747–6032

Schlather M, Malinowski A, Menck PJ, Oesting M, Strokorb K (2015) Analysis, simulation and prediction of multivariate random fields with package random fields. J Stat Softw 63(8):1–25. http://www.jstatsoft.org/v63/i08/

Slayback DA, Pinzon JE, Los SO, Tucker CJ (2003) Northern hemisphere photosynthetic trends 1982–99. Glob Chang Biol 9(1):1–15

Sobrino J, Julien Y (2011) Global trends in NDVI-derived parameters obtained from gimms data. Int J Remote Sens 32(15):4267–4279

Sobrino JA, Jiménez-Muñoz JC, Paolini L (2004) Land surface temperature retrieval from Landsat TM 5. Remote Sens Environ 90(4):434–440

Stein A, van der Meer FD, Gorte B (1999) Spatial statistics for remote sensing, vol 1. Springer Science & Business Media

Survey UG (2015) Landsat 8 (l8) data users handbook. US geological survey, Version 10(97p):1–97

Tucker CJ, Pinzon JE, Brown ME, Slayback DA, Pak EW, Mahoney R, Vermote EF, El Saleous N (2005) An extended avhrr 8-km NDVI dataset compatible with MODIS and spot vegetation NDVI data. Int J Remote Sens 26(20):4485–4498

Tüshaus J, Dubovyk O, Khamzina A, Menz G (2014) Comparison of medium spatial resolution ENVISAT-MERIS and terra-MODIS time series for vegetation decline analysis: a case study in Central Asia. Remote Sens 6(6):5238–5256

Vancutsem C, Ceccato P, Dinku T, Connor SJ (2010) Evaluation of MODIS land surface temperature data to estimate air temperature in different ecosystems over Africa. Remote Sens Environ 114(2):449–465

Venables WN, Ripley BD (2002) Modern applied statistics with S, 4th edn. Springer, New York. http://www.stats.ox.ac.uk/pub/MASS4. ISBN 0-387-95457-0

Verger A, Baret F, Weiss M, Kandasamy S, Vermote E (2013) The CACAO method for smoothing, gap filling, and characterizing seasonal anomalies in satellite time series. IEEE Trans Geosci Remote Sens 51(4):1963–1972

Viovy N, Arino O, Belward A (1992) The best index slope extraction (bise): a method for reducing noise in NDVI time-series. Int J Remote Sens 13(8):1585–1590

Wackernagel H (1995) Multivariate geostatistics: an introduction with applications. Springer Science & Business Media

Wang R, Cherkauer K, Bowling L (2016) Corn response to climate stress detected with satellite-based NDVI time series. Remote Sens 8(4):269

Geostatistical Analysis of Lead in Water

Pierre Goovaerts

Abstract The drinking water contamination crisis in Flint, Michigan has attracted national attention since extreme levels of lead were recorded following a switch in water supply that resulted in water with high chloride and no corrosion inhibitor flowing through the aging Flint water distribution system. Since Flint returned to its original source of drinking water on October 16, 2015, the State has conducted eleven bi-weekly sampling rounds, resulting in the collection of 4,120 water samples at 819 "sentinel" sites. This chapter describes the first geostatistical analysis of these data and illustrates the multiple challenges associated with modeling the space-time distribution of water lead levels across the city. Issues include sampling bias and the large nugget effect and short range of spatial autocorrelation displayed by the semivariogram. Temporal trends were modeled using linear regression with service line material, house age, poverty level, and their interaction with census tracts as independent variables. Residuals were then interpolated using kriging with three types of non-separable space-time covariance models. Cross-validation demonstrated the limited benefit of accounting for secondary information in trend models and the poor quality of predictions at unsampled sites caused by substantial fluctuations over a few hundred meters. The main benefit is to fill gaps in sampled time series for which the generalized product-sum and sum-metric models outperformed the metric model that ignores the greater variation across space relative to time (zonal anisotropy). Future research should incorporate the large database assembled through voluntary sampling as close to 20,000 data, albeit collected under non-uniform conditions, are available at a much greater sampling density.

P. Goovaerts (✉)
BioMedware, Inc, 11487 Highland Hills Drive, Jerome, MI 49249, USA
e-mail: goovaerts@biomedware.com

14.1   Introduction

The drinking water contamination crisis in Flint, Michigan has attracted national attention since extreme levels of lead were recorded in local water supplies and the percentage of children with elevated blood lead levels (BLL) increased in neighborhoods with the highest water lead levels (WLL). Problems started when the City of Flint, Michigan adopted the cost-saving decision of drawing and treating water from the Flint River instead of relying on the Detroit Water and Sewerage Department's system (DWSD) for its public water supply. A few months later, in December 2014, water samples showed elevated levels of trihaloethanes (THMs) a disinfection byproduct of chlorine, as well as high levels of lead and copper. A public health emergency was declared and residents were told to avoid drinking the water until it was tested or approved water filters were installed. In July 2015, public concerns were raised that lead and copper were being leached from corrosion (chlorine-induced) in the underground lead service lines and home plumbing fixtures as a result of not using corrosion control treatment (CCT). In August and September 2015, 16.6% of the 271 water samples collected by a Virginia Tech's team were found to exceed the EPA action level of 15 µg/L (ATSDR 2010). In September and October 2015, elevated childhood blood lead levels were confirmed and an emergency response was initiated (Hanna-Attisha et al. 2016), leading the city to switch back to the DWSD water supply on October 16, 2015.

Starting in February 2016, samples were collected bi-weekly at more than 600 sentinel sites chosen by the EPA and MDEQ (Michigan Department of Environmental Quality) across the city to determine the general health of the distribution system and to track changes in lead concentrations over time (Flint Safe Drinking Water Task Force 2016). After five rounds of sentinel sampling, a new sentinel program called "Extended Sentinel Site Program" started in June 2016, targeting specifically sites with high WLL during previous rounds or located in the highest-risk areas. Six additional sampling rounds were conducted for this smaller network including fewer than 200 sites. Overall these 11 sampling rounds resulted in the collection of 4,120 data at 819 different sites over a 40-week time period. This State-controlled monitoring program was supplemented by a voluntary or homeowner-driven sampling whereby concerned citizens received a testing kit and conducted sampling on their own (Goovaerts 2017a, b). Despite the larger size of this database (18,760 samples collected over 53 weeks at 10,341 sites), its heterogeneity and lack of systematic sampling across time prohibited its use in the present space-time analysis.

Except for a few graphs and location maps, the database assembled by the City of Flint and made available online has not undergone any rigorous statistical treatment by State employees and only a few studies have been published so far. Using a data-driven approach Abernethy et al. (2016) developed an ensemble of predictive models (e.g., random forest, logistic regression, linear discriminant analysis) to assess the risk of lead contamination in individual homes and neighborhoods in Flint. They trained these models using a wide range of data sources, including residential water tests, historical records, and city infrastructure data. Their analysis however ignored the spatial correlation among data and did not include a temporal component. A time trend analysis was conducted by Goovaerts (2017a) who used joinpoint regression to model time series of lead levels collected by the state-controlled and voluntary sampling programs. This analysis carried out at the city and ward levels still ignored the spatial correlation among data and did not provide any tax parcel-based prediction. A space-time analysis of these data should however provide important information to identify residences where high levels of lead are expected. It would also support any assessment of past and current lead exposures among the population at risk, particularly pregnant women and children.

Geostatistical techniques have been routinely used to analyze and map the spatial variability of soil and sediment lead concentrations (Goovaerts et al. 1997; Cattle et al. 2002; Solt et al. 2015), yet their application to lead in drinking water is far less common and mainly concerns groundwater quality (Siddique et al. 2012). A recent study (Wang et al. 2014) applied geographic information systems (GIS) and a hydraulic model of distribution systems to test the influences of pipe material, pipe age, water age, and other water quality parameters on lead/copper leaching in Raleigh (NC). In Symanski et al. (2004), mixed effect models were used to assess spatial fluctuations, temporal variability, and errors due to sampling and analysis for levels of disinfection by-products in water samples collected in households within the same distribution system. To the author's knowledge, the present study is however the first application of geostatistics to lead in drinking water within a distribution system.

This chapter describes a new methodology to predict lead level in tap water, accounting for WLL measurements collected in neighboring houses, housing characteristics (e.g., age of the house or presence of lead pipes), and temporal trends (e.g., decline since return to pre-crisis source of drinking water). Linear regression was used to model temporal trends at sentinel sites, accounting for the composition of service line (SL), construction year, poverty level, and census tracts as covariates. Cross-validation analysis allowed one to assess the benefit of this approach and compare the results obtained using three different types of space-time covariance models. Both the cases of predicting unsampled times at monitored locations (i.e., filling gaps in time series) and making predictions at unsampled locations were investigated.

14.2   Materials and Methods

14.2.1   Datasets

4,150 WLL measurements recorded over the period 2/20/2016-11/20/2016 were downloaded from http://www.michigan.gov/flintwater (residential testing results).

Table 14.1 Datasets available for the space-time analysis: 4,120 water lead levels measured over 11 sampling rounds. Statistics include the number of data available, the sampling period, the percentage of WLL above 15 µg/L, the mean of logtransformed concentrations, and the composition of service line that was recorded for each sentinel site (three main categories besides plastic, unknown, and other)

Sampling round	Data (n)	Sampling period	%WLL > 15 µg/L	Mean Log_{10} (µg/L)	Composition of SL		
					Lead	Galvanized	Copper
Round S1	610	2/16/2016–2/29/2016	9.51	0.487	5.90	20.66	68.20
Round S2	606	2/24/2016–3/13/2016	8.42	0.465	8.91	19.97	67.00
Round S3	654	3/15/2016–3/24/2016	8.26	0.480	11.62	19.57	63.91
Round S4	644	3/29/2016–4/5/2016	7.14	0.457	13.66	17.39	64.29
Round S5	622	4/13/2016–4/15/2016	6.43	0.427	14.31	15.27	65.43
Round X1	170	5/23/2016–6/7/2016	7.06	0.604	45.88	9.41	44.71
Round X2	178	6/14/2016–6/30/2016	8.99	0.638	49.44	7.87	42.70
Round X3	167	7/19/2016–7/22/2016	6.59	0.557	46.11	8.38	45.51
Round X4	162	8/18/2016–8/22/2016	9.88	0.579	45.06	9.26	45.68
Round X5	158	9/19/2016–9/27/2016	6.33	0.522	45.57	9.49	44.94
Round X6	149	11/17/2016–11/23/2016	6.71	0.532	45.64	9.40	44.97

Data were then allocated to an individual tax parcel unit on the basis of their postal address. Data with incomplete address (two samples) or duplicates (e.g., samples taken from two different faucets on the same day in the same house) were discarded, leading to a total of 4,120 samples collected at 819 different sites; see Table 14.1. Because of their strongly positively skewed distribution (concentrations range from 0 to 5,986 µg/L) and large proportion of zero values (34.6%), data were transformed using the following formula $Log_{10}(z+1)$.

Sentinel sites were initially selected from a pool of 1,951 volunteer sites identified during door-to-door water distribution; in particular it included all 156 sites with lead or lead combination service lines according to City records. Other sites were added according to several criteria: (i) spatial distribution to ensure coverage of all nine City wards, (ii) measurements of high blood levels (Hanna-Attisha et al. 2016), and (iii) environmental justice considerations (e.g. presence of houses with lead-based paint, minority population, and lower socio-economic households). This

Table 14.2 Statistics computed for time series of different lengths: number of sentinel sites, percentage of WLL above 15 μg/L, the mean of logtransformed concentrations, and the composition of service line

Length	Sites	%WLL > 15 μg/L	Mean Log_{10} (μg/L)	Composition of SL		
				Lead	Galvanized	Copper
1	80	7.50	0.413	6.25	22.50	66.25
2	33	6.06	0.475	12.12	18.18	63.64
3	36	6.48	0.433	22.22	25.00	46.30
4	95	4.74	0.411	5.26	18.95	70.53
5	409	3.52	0.358	2.93	17.85	73.59
6	41	8.54	0.530	21.95	4.88	73.17
7	19	9.77	0.651	89.47	5.26	5.26
8	10	11.25	0.705	68.75	7.50	23.75
9	23	11.59	0.693	84.54	4.35	11.11
10	32	18.75	0.750	38.75	15.63	45.63
11	41	19.82	0.793	33.92	20.26	45.81

initial set evolved between sampling rounds as some residents stopped participating, while others asked to be included in the network (Goovaerts 2017b), which explains the fluctuation in the number of sampled sites during the first five rounds S1-S5: 607–621 (Table 14.1). Fewer sites (149–178) were then part of the "Extended Sentinel Site Program". Table 14.2 indicates that only 41 sites were sampled in all 11 rounds, while 80% of time series included five observations or less.

Each house selected to be part of the sentinel network was visited by a licensed plumber who classified the material of the service line coming into the home (i.e., customer-side service line) into six categories: lead, galvanized, copper, plastic, other, and unknown. Galvanized refers to iron pipe with a protective "galvanized" surface coating composed of zinc, lead, and cadmium, and therefore can be a long-term source of lead (Clark et al. 2015). The term "unknown" was used whenever the SL material could not be confirmed because, for example, the line was behind a wall or way back in a crawl space.

City records were the only source of service line data available for the majority of 56,039 tax parcels which were not part of the sentinel sampling program. These records are however inaccurate and lead to the over-identification of lead SLs, likely because old records were not updated as these lines were being replaced (Goovaerts 2017c). The same author found that construction year was a good predictor of service line material: galvanized lines were mostly found in pre-1934 houses, while the frequency of lead service lines (LSLs) peaked for houses built around World War II. This information was combined with field inspection data and city records to predict by indicator kriging the likelihood that a home has lead or galvanized SL (Goovaerts 2017c).

Besides service lines, lead in drinking water mainly comes from lead-based solder and lead-containing plumbing fixtures (Lee et al. 1989; Cartier et al. 2011). Plumbing material is usually related to the installation year of a plumbing system, which can be approximated by the year of construction. For example, most faucets purchased prior to 1997 were made of brass or chrome-plated brass containing up to 8 percent lead (Rabin 2008). Construction year was retrieved from the 2016 Parcels GIS layer. The attribute "Year_built" was missing for 20,372 parcels and was estimated by ordinary kriging (Goovaerts 1997) with a mean absolute error of prediction of 6.43 years. Based on its relationship to water lead levels (Goovaerts 2017a), construction year was discretized into three classes: pre-1940, 1940–1959, and post-1959.

Poor workmanship as well as lack of regular maintenance can also lead to more corrosion and leaching, and the presence of lead particulates, such as disintegrating brass or detaching pieces of old solder (Wang et al. 2014). Socio-economic status was here assessed using 2015 ACS (American Community Survey) 5-year estimates of the percentage of the block group population living in households where the income is less than or equal to twice the federal "poverty level".

There are many other variables known to influence lead in drinking water. For example, longer water age (i.e., water travel time between the treatment plant and home plumbing system) can decrease the effectiveness of corrosion control; increasing leaching and water lead levels (US EPA 2002; Wang et al. 2014). This information was however unavailable for this study.

14.2.2   Space-Time Kriging and Covariance Models

Let $z(\mathbf{u}_\alpha;t)$ denote the water lead level recorded on time t at sentinel site α georeferenced by the geographical coordinates $\mathbf{u}_\alpha = (x_\alpha, y_\alpha)$ of the corresponding tax parcel centroid. Prediction of z-value at unsampled time t_0 and location u_0 was conducted using the following kriging estimator:

$$Z^*(u_0;t_0) = \sum_{t=t_0-\Delta t}^{t_0+\Delta t} \sum_{\alpha=1}^{n(t)} \lambda_{\alpha t} \times z(u_\alpha;t) \qquad (14.1)$$

$n(t)$ is the number of observations recorded at time t, within the time window $2\Delta t$, that were retained for estimation. The weights $\lambda_{\alpha t}$ are solution of the following space-time (ST) kriging system:

$$\sum_{t=t_0-\Delta t}^{t_0+\Delta t}\sum_{\alpha=1}^{n(t)}\lambda_{\alpha t}C\left(u_\alpha-u_\beta;t-t'\right)+\mu=C\left(u_0-u_\beta;t_0-t'\right)\quad \beta=1,\cdots,n(t')$$

$$\sum_{t=t_0-\Delta t}^{t_0+\Delta t}\sum_{\alpha=1}^{n(t)}\lambda_{\alpha t}=1 \qquad\qquad t'=t_0-\Delta t,\cdots,t_0+\Delta t$$

$$(14.2)$$

The parameter μ is a Lagrange multiplier accounting for the constraint on the weights. The term $C\left(u_\alpha-u_\beta;t-t'\right)$ is the ST covariance between any two observations recorded at locations u_α and u_β at times t and t', respectively. Euclidian distances were used here since most lead in drinking water comes from premise plumbing materials and service lines instead of being transported through water mains (Del Toral et al. 2013; EET Inc. 2015).

One challenge associated with the application of ST kriging is the choice of a ST covariance model within the ever growing class of models (Montero et al. 2015). The following three non-separable ST covariance models were compared in the present study:

- The generalized product-sum model (De Iaco et al. 2002):

$$C(h,\tau)=k_1C_s(h)+k_2C_t(\tau)+k_3C_s(h)C_t(\tau) \qquad (14.3)$$

where k_1,k_2, and k_3 are non-negative (strictly positive for k_3) coefficients estimated from the sills of the spatial, temporal, and spatio-temporal semivariograms (De Cesare et al. 2002).
- The metric model (Dimitrakopoulos and Luo 1994):

$$C(h,\tau)=C_{st}\left(\sqrt{\left(\frac{h}{a_s}\right)^2+\left(\frac{\tau}{a_t}\right)^2}\right) \qquad (14.4)$$

where a normalized space-time distance measure is created by rescaling the spatial and temporal lags, h and τ, by the ranges of the spatial and temporal semivariograms, a_s and a_t (case of geometric anisotropy).
- The sum-metric model (Heuvelink and Griffith 2010):

$$C(h,\tau)=C_s(h)+C_t(\tau)+C_{st}\left(\sqrt{\left(\frac{h}{a_s}\right)^2+\left(\frac{\tau}{a_t}\right)^2}\right) \qquad (14.5)$$

This model combines characteristics of the two previous models: (i) sum of spatial and temporal covariances allowing for the presence of zonal anisotropies (i.e., semivariogram sills are not the same in all directions), and (ii) a metric ST model for the residual variability (geometric anisotropy).

Two other classes of non-separable ST covariance models, Cressie-Huang model (Cressie and Huang 1999) and Gneiting models (Gneiting 2002), were not considered because: (1) the fitting of these models needs a complex iterative parameter optimization technique (De Iaco 2010), whereas the three selected models can be fitted using straightforward techniques similar to those already used for spatial-only and temporal-only semivariograms, and (2) recent studies (Guo et al. 2015) indicate that these two more complex models provide similar fits to experimental ST semivariograms and comparable prediction accuracy as the product-sum model, confirming previous findings (De Iaco 2010).

The main difficulty in the practical implementation of the product-sum and sum-metric models is the inference of the sill of the ST semivariogram model, $C_{st}(0)$, which is most often estimated visually from the 3D plot of the experimental ST semivariogram $\hat{\gamma}_{st}(h, \tau)$ (e.g., De Cesare et al. 2002; Heuvelink and Griffith 2010). In order to make the fitting procedure more user-friendly, the space-time sill $C_{st}(0)$ was here computed as the following weighted average of experimental space-time semivariogram values:

$$C_{st}(0) = \frac{1}{\sum_h \sum_\tau w_{h,\tau}} \sum_h \sum_\tau w_{h,\tau} \hat{\gamma}_{st}(h, \tau) \quad \text{if } \hat{\gamma}_{st}(h, \tau) \geq g_c \qquad (14.6)$$

where the weight $w_{h,\tau}$ is the number of data pairs falling into the class of spatial and temporal lags (h, τ). Only the classes where the ST semivariogram values exceed a critical sill g_c, defined as the maximum of the spatial and temporal sills, were used.

14.2.3 Accounting for Secondary Information

Lead service lines are widely considered the main source of lead in drinking water (Lee et al. 1989; Clark et al. 2015). Another culprit is lead fixtures and pipes present within old houses (premises plumbing), and poverty can compound the problems through the lack of maintenance. Goovaerts (2017a) also found that temporal trends can vary greatly across the city. This secondary information was here incorporated in the definition of a stochastic trend model $M(u;t)$, leading to the following decomposition of the space-time random function (RF) (Kyriakidis and Journel 1999):

$$Z(u;t) = M(u;t) + R(u;t) \qquad (14.7)$$

where $M(u;t)$ is a nonstationary spatiotemporal RF modeling the space-time distribution of the mean process, with $E[M(u;t)] = m(u;t)$ and $R(u;t)$ is a zero mean stationary spatiotemporal RF modeling space-time fluctuations around $M(u;t)$.

The trend component at each sentinel site u_α was fitted using a linear model including six fixed factors: presence/absence of LSL, presence/absence of galvanized service line (GSL), time since first sample was collected (TIME), poverty level (POV), house age (AGE), and census tract (CT). The model takes the following form:

$$
\begin{aligned}
M(u;t) = {} & LSL(u) \times TIME + CT(u) \times TIME + LSL(u) \times CT(u) \\
& + GSL(u) \times CT(u) + AGE(u) \times CT(u) \\
& + POV(u) \times CT(u)
\end{aligned}
\tag{14.8}
$$

This model naturally handles uneven spacing of repeated measurements within each time series, as well as their correlation which was modeled using a spherical variance-covariance structure. Once the trend model was fitted, regression residuals were interpolated using space-time simple kriging and the ST covariance models introduced in Sect. 14.2.2.

14.2.4 Cross-Validation

The accuracy of the predictive models created by the different approaches (e.g., three types of ST covariance models, univariate vs incorporation of secondary information) was assessed by cross-validation whereby each observation or time series (i.e., all data collected at the same site) was removed at a time and re-estimated using data collected at neighboring sentinel sites. The following performance criteria were then computed from n kriging estimates:

- the mean error (ME) of prediction as:

$$
ME = \frac{1}{n} \sum_{t=1}^{T} \sum_{a=1}^{n(t)} \left(z^*(u_a;t) - z(u_a;t) \right)
\tag{14.9}
$$

- the mean absolute error (MAE) of prediction as:

$$
MAE = \frac{1}{n} \sum_{t=1}^{T} \sum_{a=1}^{n(t)} \left| z^*(u_a;t) - z(u_a;t) \right|
\tag{14.10}
$$

- the mean square standardized residual (MSSR) as:

$$
MSSR = \frac{1}{n} \sum_{t=1}^{T} \sum_{a=1}^{n(t)} \frac{\left(z^*(u_a;t) - z(u_a;t) \right)^2}{\sigma_K^2(u_a;t)}
\tag{14.11}
$$

where $\sigma_K^2(u_a;t)$ is the kriging variance.

A mean error close to zero indicates a lack of bias, while the mean absolute error should be as small as possible. If the actual estimation error is equal, on average, to the error predicted by the model, the MSSR statistic should be about one (Wackernagel 1998, p. 91).

One application of the predictive models is to prioritize any further sampling or intervention by ranking tax parcels from highly hazardous to less hazardous on the basis of kriging estimates. The ability of this ranking to identify successfully sites where WLL is greater or equal to the EPA action level of 15 µg/L was assessed using Receiver Operating Characteristics (ROC) curves which plot the probability of false positive versus the probability of detection (Swets 1988; Fawcett 2006; Goovaerts et al. 2016). The accuracy of the classification was quantified using the relative area under the ROC curve (AUC statistic), which ranges from 0 (worst case) to 1 (best case). The AUC is equivalent to the probability that the classifier will rank a randomly chosen positive instance (e.g., $z_c \geq 15$ µg/L) higher than a randomly chosen negative instance (e.g., $z_c < 15$ µg/L).

14.3 Results and Discussion

14.3.1 Spatial Distribution

Figure 14.1a shows the location of all 819 sentinel sites within the nine wards in the city of Flint. Site-specific statistics such as number of observations and average log concentrations recorded for each time series, as well as composition of service line (GSL vs. LSL), were aggregated at the census tract level for better visualization. Geographical clusters of sentinel sites can be distinguished in several census tracts (e.g. border of wards 2 and 6, wards 7 and 9) which tend to be the tracts with the largest WLLs (Fig. 14.1c) and percentages of sampled LSLs (Fig. 14.1d). There is also a clear spatial trend with fewer lead service lines (e.g., none in Ward 1) and shorter time series (Fig. 14.1b) sampled in the Northern part of the city. Ward 5 includes the oldest neighborhood where GSLs are prevalent (Fig. 14.1e), while LSLs appear as small clusters, in particular in wards 6, 7 and 9 (Goovaerts 2017c).

14.3.2 Temporal Trend Modeling

Temporal trends for the three major types of service line were visualized by aggregating observations within non-overlapping 14-day windows, which corresponds to the average time interval between sampling rounds during the first phase (Round S) of the sentinel monitoring program (Table 14.1). Except for LSLs water lead levels do not appear to have declined over the 40-week sampling period; actually they seem to have slightly increased for GSLs (Fig. 14.2a). These results are however a direct artifact of the sampling strategy whereby 80% of sentinel sites

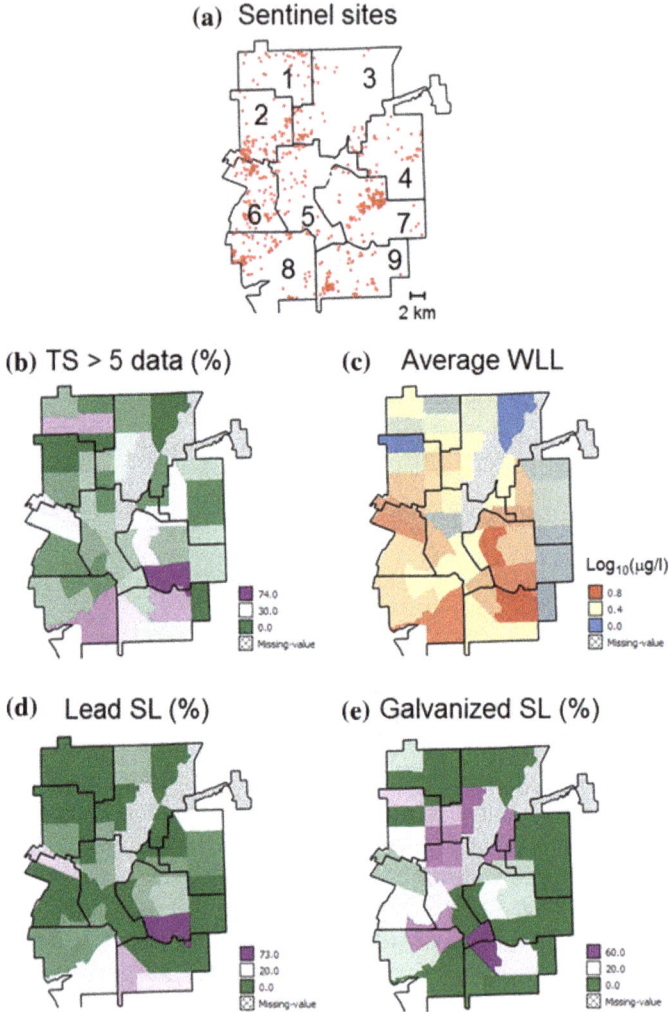

Fig. 14.1 a Location of sentinel sites in each of the nine wards, and several census tract-level statistics: **b** percentages of time series (TS) including more than five observations, **c** average water lead levels, **d** percentage of sites with lead service lines, **e** percentage of sites with galvanized service lines. Shaded polygons indicate census tracts that do not include any sentinel site (missing values)

were not sampled beyond week 16, while sampling continued at sites where the risk of exceeding the EPA action level of 15 µg/L was the greatest (Table 14.2).

After elimination of all sites where fewer than six observations were collected, the averaged time series display the expected decline (Fig. 14.2b). The impact is minimal for LSLs since most of these sites are considered at risk and were sampled during both the initial and extended sentinel sampling programs (Rounds S and X).

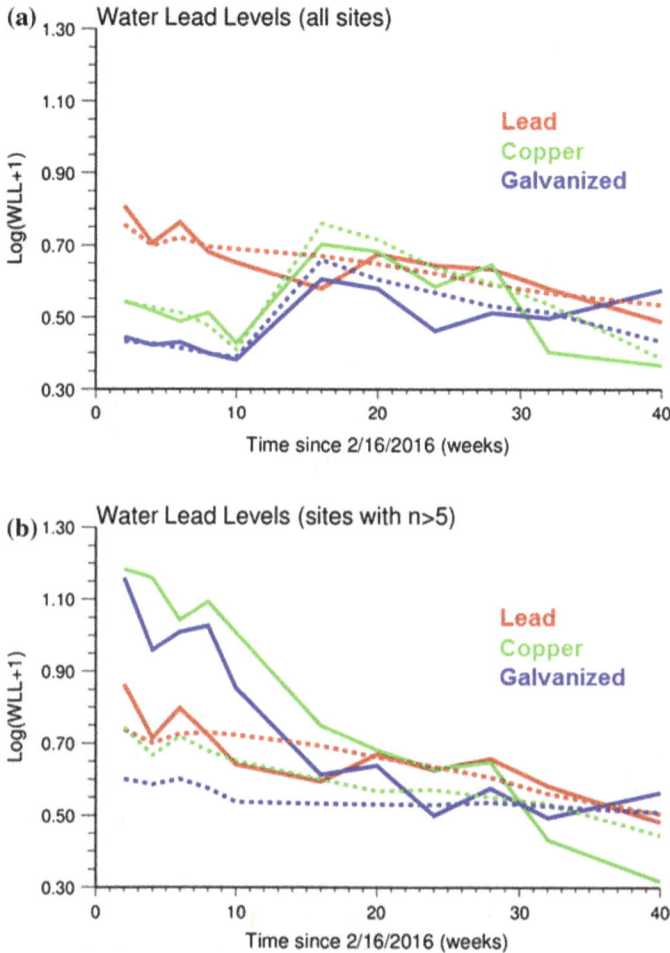

Fig. 14.2 Time series of observed (solid line) and predicted by regression (dashed line) water lead levels computed on average for the three major types of service line: lead, galvanized, and copper. Results (log transformed concentrations) are calculated from: **a** all sites, and **b** subset of sites where at least six observations were recorded

The selection bias is stronger for copper and galvanized lines, which explains the larger water lead levels recorded during the first 16 weeks relative to LSLs.

This sampling bias complicated greatly the modeling of temporal trends by regression. Indeed using all the data would underestimate the weekly rate of decline of water lead levels, whereas subsetting the dataset (e.g., using only time series including more than five data points as in Fig. 14.2b) will result in overestimating the concentrations at a majority of sites. In addition, the time series length cannot be used as covariate in the model to allow its application at unmonitored locations. Two modeling strategies were considered in this chapter. First, because of its relationship with time series length (Fig. 14.1) census tract was used as covariate in the regression model (Eq. 14.8). The second more complicated approach was to allow the intercept to fluctuate among sentinel sites, even when located within the same tract; i.e., use a mixed model where the intercept is modeled as a random effect. The trade-off cost for this added flexibility was the need to estimate the intercept at unmonitored locations, which was accomplished using ordinary kriging. Despite providing a better fit than the first alternative, the mixed model did not lead to more accurate kriging estimates, hence only the first option is discussed hereafter.

All six interaction terms in the trend model (Eq. 14.8) were highly significant ($\alpha = 0.01$). The correlation between predicted and observed WLL is however rather weak ($r = 0.47$), which illustrates the challenge of predicting spatial and temporal variations in lead for drinking water (Bailey and Russell 1981; Del Toral et al. 2013). While the output of the regression model provides a reasonable fit to the SL-specific time series computed using all the data (Fig. 14.2a), it underestimates water lead levels for LSL and GSL when using only time series including more than five data points (Fig. 14.2b).

14.3.3 Variography

Semivariograms helped quantifying the scale and magnitude of the space-time variability displayed by the maps and time series of Figs. 14.1 and 14.2. The spatial semivariogram (Fig. 14.3a) shows three nested scales of spatial variability: (1) a long range (2.35 km) caused by the neighborhood effect since houses in the same neighborhood tend to be built at the same time (i.e., similar plumbing system) and have similar water age, (2) a short range (200 m) corresponding to variability between adjacent houses, and (3) a nugget effect or discontinuity at the origin which represents the variability among samples taken within the same tax parcel (i.e. different apartments and/or measurement error for samples taken within the same residence). The substantial short-range variability (71% of total sill) likely reflects the heterogeneity in housing conditions (e.g., renovated houses) as well as the lack of uniformity of sampling conducted by homeowners since even with simple instructions it is difficult to ensure strict adherence to any sampling protocol (Del Toral et al. 2013). This interpretation is confirmed by the similar short-range variability displayed by the semivariogram of regression residuals (Fig. 14.3a, lower blue curve) since the regression model (Eq. 14.8) does not account for sampling characteristics. It is noteworthy that the longer range of 2.35 km is still fairly small relative to the size of the city (see legend of Fig. 14.1a), while the average separation distance between each sentinel site and the closest neighbor (293 m) exceeds the shortest range (200 m) that encapsulates 71% of the total spatial variability.

The temporal semivariogram (Fig. 14.3b) also displays three nested scales of variability although the longer range structure (110 days) represents here 53% of the total variability. Another difference with the spatial case is the overlap of

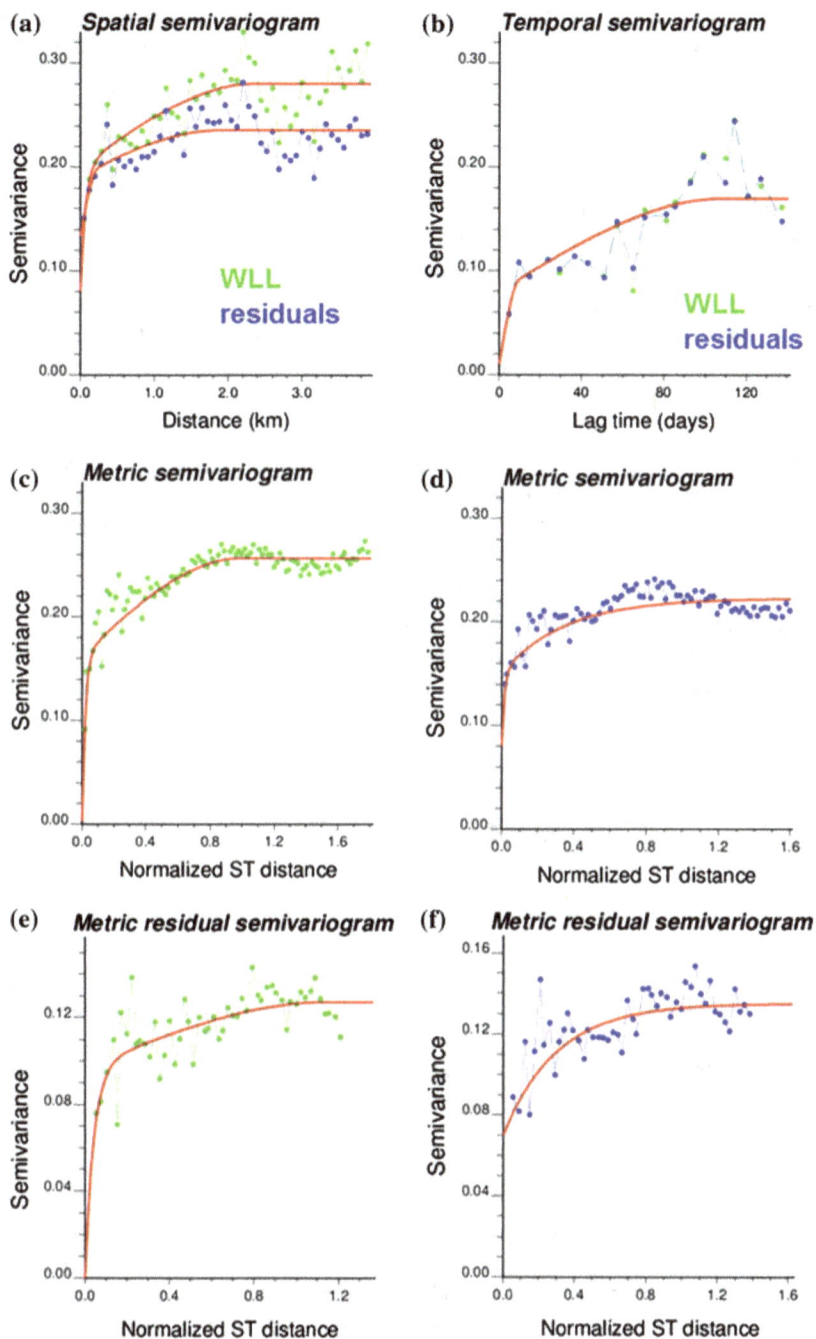

(a) **Spatial semivariogram**

(b) **Temporal semivariogram**

(c) **Metric semivariogram**

(d) **Metric semivariogram**

(e) **Metric residual semivariogram**

(f) **Metric residual semivariogram**

◀Fig. 14.3 Experimental semivariograms with the model fitted that were used to form the three types of ST covariance models (Eqs. 14.3–14.5) **a** spatial semivariogram (lower curve is for residuals), **b** temporal semivariogram, **c** metric semivariogram for WLLs, **d** metric semivariogram for regression residuals, **e** metric residual semivariogram (sum-metric model) for WLLs, **f** metric residual semivariogram for regression residuals

temporal semivariograms for WLLs and regression residuals, illustrating the inability of the trend model (Eq. 14.8) to capture purely temporal changes. This result is in agreement with the small magnitude of changes displayed by the time series of predicted values in Fig. 14.2 (dashed line). Comparison of the total sills of spatial and temporal semivariograms (Fig. 14.3a–b) indicates that the variability observed across space is greater than the temporal variability. Such zonal aniso-tropy is in conflict with the assumption underlying the metric ST covariance model (Eq. 14.4).

Figure 14.3c–d show the semivariograms computed using a normalized space-time distance (metric model). Because the spatial and temporal lags were rescaled using different constants for the WLL and residual semivariograms, these two curves are plotted separately. The vertical axis is however comparable and illustrates the smaller variability of residuals (i.e., lower sill for the semivariogram of Fig. 14.3d). Once again, both semivariograms display substantial short-range variability. The last two semivariograms (Fig. 14.3e–f) represent the metric space-time model that captures the residual variability in the sum-metric model (Eq. 14.5).

14.3.4 Cross-Validation Analysis

The semivariogram models of Fig. 14.3 were used to conduct a cross-validation analysis whereby one observation (LOO approach) or one time series (LTO approach) was removed at a time and re-estimated using data collected at neigh-boring sentinel sites. Based on a sensitivity analysis using ST ordinary kriging and MAE criterion, 48 observations with a maximum of three data points per site were retained for the estimation by univariate and residual ST kriging. Results obtained for predictions by the time trend model were also included as reference in Table 14.3.

The first three rows in Table 14.3 indicate that all algorithms give unbiased predictions (ME close to zero). As expected, the best prediction scores (i.e., lower MAE and higher AUC) are obtained when using data from the same time series (LOO approach) instead of relying solely on non-colocated data (LTO approach). Except for MSSR the product-sum model performs best, with the sum-metric model being a close second. The metric model underperforms the other two models because the combination of both spatial and temporal dimensions through a nor-malized space-time distance leads one to underestimate the correlation among observations of the same time series. In other words, the assumption underlying the

Table 14.3 Results of cross-validation analysis conducted by leaving one observation out (LOO) or one time series out (LTO) at a time. The four performance criteria described in Sect. 14.2.4 were computed for three types of space-time covariance models (generalized product-sum, metric, and sum-metric) and three space-time interpolation algorithms (ST ordinary kriging, trend model fitted by linear regression with and without interpolation by ST residual kriging)

Algorithm	Performance criteria					
	Product-sum model		Metric model		Sum-metric model	
	LOO	LTO	LOO	LTO	LOO	LTO
Mean error of prediction (ME)						
ST ordinary kriging	−0.001	0.009	0.003	0.007	−0.001	0.008
ST residual kriging	0.0	0.008	0.003	0.005	0.001	0.008
Trend model[a]	0.0					
Mean absolute error of prediction (MAE)						
ST ordinary kriging	0.257	0.375	0.336	0.384	0.263	0.378
ST residual kriging	0.251	0.337	0.318	0.346	0.254	0.343
Trend model[a]	0.331					
Mean square standardized residual (MSSR)						
ST ordinary kriging	1.326	0.954	1.026	1.208	1.190	1.111
ST residual kriging	1.119	0.912	0.957	1.086	1.015	1.086
Trend model[a]	74.9					
Area under the ROC curve for 15 µg/L (AUC)						
ST ordinary kriging	0.832	0.615	0.743	0.598	0.829	0.613
ST residual kriging	0.839	0.707	0.768	0.692	0.836	0.697
Trend model[a]	0.713					

[a]value for trend model is the same for all six combinations

metric model is incompatible with the zonal anisotropy detected on Fig. 14.3. Accounting for secondary information through residual kriging slightly improves the prediction relative to ST ordinary kriging; both kriging algorithms outperformed the trend model.

These results however apply only to the narrow situation where exposure to lead in drinking water is reconstructed at the sole sentinel sites. For prediction at sites where no data was collected, LTO results indicate that differences between ST covariance models are much smaller as purely temporal correlations are not used in the kriging system. Nevertheless, the product-sum model still performs best. The LTO approach also emphasizes the benefit of using trend models that account for secondary information (i.e., larger differences between residual kriging and ordinary kriging). Yet, prediction performances actually deteriorate when kriged residuals are added to the trend model: the sole trend model gives better prediction than residual kriging. It is however noteworthy that the trend model was not cross-validated, hence the observation being predicted was used to create the model.

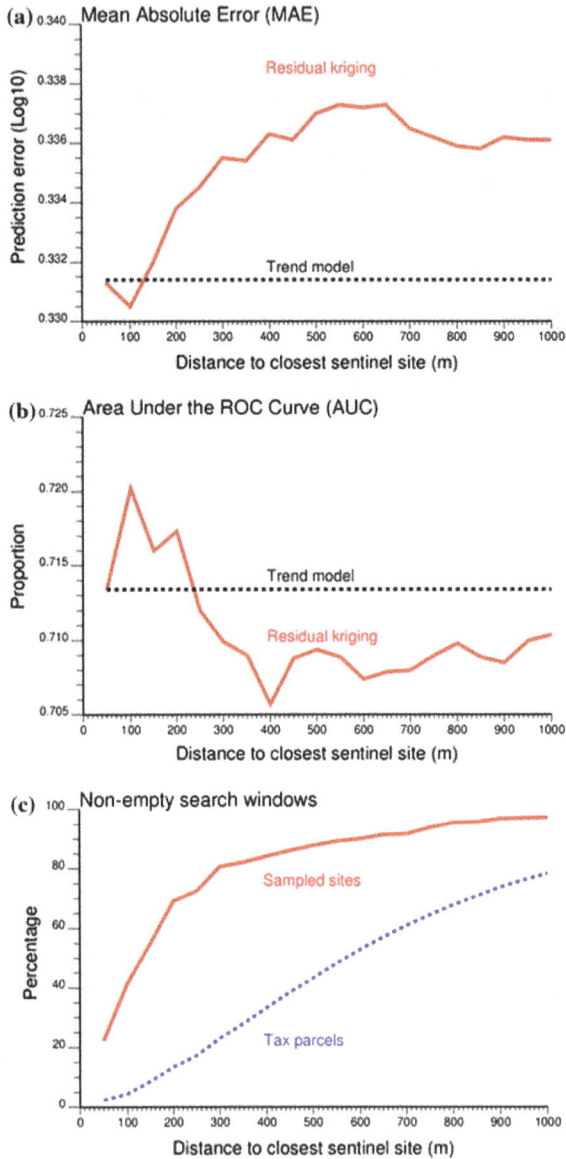

Fig. 14.4 Impact of the size of kriging search window on several statistics computed by the leave one time series out (LTO) approach: **a** mean absolute error of prediction, and **b** area under the ROC curve. Horizontal dashed lines represent the values obtained for the time trend model created by linear regression. **c** percentages of search windows that include at least one observation when centered on sampled sentinel sites or tax parcels

Because of the substantial short-scale spatial variability retaining increasingly distant data is expected to add more and more noise to the kriging estimate. This was investigated by changing the search strategy and selecting only sentinel sites located within a given distance of the site being predicted. If no data was located within the search radius, the kriged residual was zero and the residual kriging estimate was simply the value of the trend model. Figure 14.4 shows results of this sensitivity analysis conducted for the product-sum model over distances ranging from 50 m to 1 km. For the mean error of prediction the little benefit of residual kriging vanishes as soon as data beyond 100 m are used in the estimation (Fig. 14.4a), while this distance is 200 m for the area under the ROC curve (Fig. 14.4b). Figure 14.4c indicates that 42% of sentinel sites have another sentinel site within 100 m, while this percentage is only 4.6% for tax parcels (Fig. 14.4c). In other words, there is little benefit in applying geostatistics to model the space-time distribution of WLL over the 56,039 tax parcels in Flint using the data collected at sentinel sites.

14.4   Conclusions

This chapter presented the first application of space-time geostatistics to lead levels recorded in drinking water of a public distribution system. The methodology was illustrated using 4,120 water samples that were collected at 819 "sentinel" sites over a 40-week period in the city of Flint. Despite a sizable database assembled by the State of Michigan, the geostatistical analysis was hampered by a temporal sampling bias and the existence of substantial variability over a few hundred meters. Unlike other countries such as Canada or France, sampling is not conducted by a trained technician in the US. Instead, homeowners are expected to collect water samples after a minimum of 6 h. of stagnation (e.g., overnight stagnation) following specific instructions (US EPA 2016), which can cause substantial variability among households. Other sources of fluctuation include heterogeneity in the plumbing system (e.g., renovation, installation of a new meter), location of sampled faucets (e.g., bathroom vs. kitchen), or water temperature (e.g., lead solubility increases with water temperature), to name a few.

In the present case-study, space-time kriging proved beneficial only in the situation where observations had been collected at the site being predicted; i.e., to fill the gaps in time series. The generalized product-sum and sum-metric space-time covariance models then outperformed the metric model that ignores the greater variation across space relative to time (zonal anisotropy). Sentinel sites represent however only 1.5% of tax parcels in the city of Flint. At unsampled sites the kriging prediction was no better than the temporal trend estimated by linear regression and it turned out to become less accurate if no data was collected within 100 meters. Although the regression model included site-specific characteristics, such as construction year and composition of service lines, it was unable to explain the short-range variability, leaving 78% of the total variance unaccounted for ($R^2 = 22\%$).

In the future, several approaches will be investigated to tackle the impact of short-range variability on prediction. First, the data analyzed in this chapter represent less than 20% of the water samples available for the city of Flint. The majority of samples were collected by voluntary sampling whereby concerned citizens received a testing kit and conducted sampling on their own (Goovaerts 2017a, b). Despite the lack of periodic sampling in time and existence of temporal bias (e.g., houses with low lead levels were less likely to be tested again) the greater spatial coverage (i.e., more than 18% of tax parcels sampled) will reduce substantially the average distance between a tax parcel and the closest observation. However, spatial heterogeneity will likely still be present over short distances, leading one to question our ability to make prediction at the tax parcel level. More appropriate spatial supports for prediction could be census block groups which are statistical divisions of census tracts and are generally defined to contain between 600 and 3,000 people. The city of Flint includes 132 block groups and 40 census tracts. Such spatial aggregation or upscaling would be a way to filter between-household fluctuations which appears to be mainly noise. As more US cities are facing similar drinking water crisis, reliable techniques for sampling and modeling spatial and temporal changes in water lead levels will be sorely needed.

Acknowledgements This research was funded by grant R44 ES022113-02 from the National Institute of Environmental Health Sciences. The views stated in this publication are those of the author and do not necessarily represent the official views of the NIEHS.

References

Abernethy J, Anderson C, Dai C et al (2016) Flint water crisis: data-driven risk assessment via residential water testing. arXiv:1610.00580. https://arxiv.org/abs/1610.00580. Accessed 26 May 2017

Agency for Toxic Substances and Disease Registry (2010) Case studies in environmental medicine (CSEM) lead toxicity. Course: WB 1105 Original Date August 15. US Department of Health and Human Services, Public Health Service: Atlanta

Bailey RJ, Russell PF (1981) Predicting drinking water lead levels. Environ Technol Lett 2:57–66. https://doi.org/10.1080/09593338109384023

Cartier C, Laroche L, Deshommes E et al (2011) Investigating dissolved lead at the tap using various sampling protocols. J Am Water Works Assoc 103:55–67

Cattle JA, McBratney AB, Minasny B (2002) Kriging method evaluation for assessing the spatial distribution of urban soil lead contamination. J Environ Qual 31:1576–1588

Clark BN, Masters SV, Edwards MA (2015) Lead release to drinking water from galvanized steel pipe coatings. Environ Eng Sci 32:713–721. https://doi.org/10.1089/ees.2015.0073

Cressie N, Huang H-C (1999) Classes of nonseparable, spatio-temporal stationary covariance functions. J Am Stat Assoc 94:1330–1340

De Cesare L, Myers DE, Posa D (2002) FORTRAN programs for space–time modeling. Comput Geosc 28:205–212

De Iaco S (2010) Space–time correlation analysis: a comparative study. J Appl Stat 37:1027–1041

De Iaco S, Myers DE, Posa D (2002) Nonseparable space-time covariance models: some parametric families. Math Geol 34:23–42

Del Toral MA, Porter A, Schock MR (2013) Detection and evaluation of elevated lead release from service lines: a field study. Environ Sci Technol 47:9300–9307

Dimitrakopoulos R, Luo X (1994) Spatiotemporal modeling: covariances and ordinary kriging systems. In: Dimitrakopoulos R (ed) Geostatistics for the next century. Kluwer, Dordrecht, pp 88–93

EET Inc (2015) Evaluation of lead sampling strategies. WRF, Final grant report 4569. http://www.waterrf.org/PublicReportLibrary/4569.pdf. Accessed 26 May 2017

Fawcett T (2006) An introduction to ROC analysis. Pattern Recogn Lett 27:861–874

Flint Safe Drinking Water Task Force Recommendations on MDEQ's Draft Sentinel Site Selection (2016). https://www.epa.gov/sites/production/files/2016-02/documents/task_force_recommendations_on_sentinel_site_selection_2-16.pdf. Accessed 26 May 2017

Goovaerts P (1997) Geostatistics for natural resources evaluation. Oxford University Press, New-York

Goovaerts P (2017a) The drinking water contamination crisis in Flint: modeling temporal trends of lead level since returning to Detroit water system. Sci Total Environ 581–582:66–79

Goovaerts P (2017b) Monitoring the aftermath of Flint drinking water contamination crisis: another case of sampling bias? Sci Total Environ 590–591:139–153

Goovaerts P (2017c) How geostatistics can help you find lead and galvanized water service lines: the case of Flint, MI. Sci Total Environ 599–600:1552–1563

Goovaerts P, Webster R, Dubois J-P (1997) Assessing the risk of soil contamination in the Swiss Jura using indicator geostatistics. Environ Ecol Stat 4:31–48

Goovaerts P, Wobus C, Jones R et al (2016) Geospatial estimation of the impact of deepwater horizon oil spill on plant oiling along the Louisiana shorelines. J Environ Manag 180:264–271

Gneiting T (2002) Nonseparable, stationary covariance functions for space-time data. J Am Stat Assoc 97:590–600

Guo L, Lei L, Zeng Z et al (2015) Evaluation of spatio-temporal variogram models for mapping Xco2 using satellite observations: a case study in China. IEEE J Sel Topics Appl Earth Observ Remote Sens 8:376–385

Hanna-Attisha M, LaChance J, Sadler RC et al (2016) Elevated blood lead levels in children associated with the Flint drinking water crisis: a spatial analysis of risk and public health response. Am J Public Health 106:283–290

Heuvelink GBM, Griffith DA (2010) Space–time geostatistics for geography: a case study of radiation monitoring across parts of Germany. Geogr Anal 42:161–179

Kyriakidis PC, Journel AG (1999) Geostatistical space-time models: a review. Math Geol 31:651–684

Lee RG, William CB, David WC (1989) Lead at the tap: sources and control. J Am Water Works Ass 81:52–62

Montero JM, Fernandez-Aviles G, Mateu J (2015) Spatial and spatio-temporal geostatistical modeling and kriging. Wiley, New York

Rabin R (2008) The lead industry and lead water pipes "A Modest Campaign". Am J Public Health 98:1584–1592

Siddique A, Zaigham NA, Mohiuddin S et al (2012) Risk zone mapping of lead pollution in urban groundwater. J Basic Appl Sci 8:91–96

Solt MJ, Deocampo DM, Norris M (2015) Spatial distribution of lead in Sacramento, California, USA. Int J Environ Res Public Health 12:3174–3187

Symanski E, Savitz DA, Singer PC (2004) Assessing spatial fluctuations, temporal variability, and measurement error in estimated levels of disinfection by-products in tap water: implications for exposure assessment. Occup Environ Med 61:65–72

Swets JA (1988) Measuring the accuracy of diagnostic systems. Science 240:1285–1293

US Environmental Protection Agency, Office of Ground Water & Drinking Water (2002) Effect of water age on distribution system water quality. https://www.epa.gov/sites/production/

files/2015-09/documents/2007_05_18_disinfection_tcr_whitepaper_tcr_waterdistribution.pdf. Accessed 26 May 2017

US Environmental Protection Agency, Office of Ground Water & Drinking Water (2016) Memorandum: clarification of recommended tap sampling procedures for purposes of the lead and copper rule. https://www.epa.gov/sites/production/files/2016-02/documents/epa_lcr_sampling_memorandum_dated_february_29_2016_508.pdf. Accessed 26 May 2017

Wackernagel H (1998) Multivariate geostatistics, 2nd completely revised edition. Springer, Berlin

Wang Z, Devine H, Zhang W et al (2014) Using a GIS and GIS-assisted water quality model to analyze the deterministic factors for lead and copper corrosion in drinking water distribution systems. J Environ Eng 140:A4014004

15

Hyperspectral Remotely Sensed Satellite Data and the Application of SPM

Sean A. McKenna

Abstract Spatial fields represent a common representation of continuous geoscience and environmental variables. Examples include permeability, porosity, mineral content, contaminant levels, seismic impedance, elevation, and reflectance/ absorption in satellite imagery. Identifying differences between spatial fields is often of interest as those differences may represent key indicators of change. Defining a significant difference is often problem specific, but generally includes some measure of both the magnitude and the spatial extent of the difference. This chapter demonstrates a set of techniques available for the detection of anomalies in difference maps represented as multivariate spatial fields. The multiGaussian model is used as a model of spatially distributed error and several techniques based on the Euler characteristic are employed to define the significance of the number and size of excursion sets in the truncated multiGaussian field. This review draws heavily on developments made in the field of functional magnetic resonance imaging (fMRI) and applies them to several examples motivated by environmental and geoscience problems.

15.1 Introduction

A general problem in geological and environmental investigations is rapid and accurate identification of anomalous measurements from one, two or three-dimensional data. Example applications include cluster identification in spatial point processes (e.g., Byers and Raftery 1998; Cressie and Collins 2001) detection of anomalies in remotely sensed imagery (e.g., Stein et al. 2002) and identification of anomalous clusters in lattice data (e.g. Goovaerts 2009).

S. A. McKenna (✉)
IBM Research, Dublin, Ireland
e-mail: seanmcke@ie.ibm.com

The problem of anomaly detection is complicated when the data set is composed of more than a handful of variables (multi-variate) and becomes even more complex when the multiple variables comprise a random field exhibiting spatial correlation.

The temporal and/or spatial correlation of the data rules out the application of standard statistical tests for change detection and has also limited the development of hypothesis testing techniques for correlated data (Gilbert 1987). For applications with correlated data, simulation techniques can often be used to develop the null distribution, but development of closed form hypothesis tests for analysis of the spatial random fields associated with geostatistics has remained sparse.

One approach to detection of anomalies in spatially correlated data are Local Indicators of Spatial Association (LISA) statistics (Anselin 1995; Goovaerts et al. 2005; Goovaerts 2009). These tests focus on the local relationships between adjacent cells and explore combinations of cells defined with an adjacency matrix and or a moving window visiting all cells in a lattice. A very different approach is to model the difference between images as a continuous random field and use properties of an underlying random field model to identify anomalies.

Change detection in spatial-temporal data sets has received considerable attention over the past 15–20 years within the medical imaging research community (Brett et al. 2003; Friston et al. 1994, 1995; Worsley et al. 1992, 1996) and a significant development of this research has been Statistical Parametric Mapping (SPM).

The practice of statistical parametric mapping has been developed in the field of medical imaging, particularly in brain imaging, and in the practice of functional magnetic resonance imaging (fMRI) of the brain while the subject is performing various tasks (functions). Friston et al. (1995, p. 190) provide a concise definition of SPM: "*one proceeds by analyzing each voxel using any (univariate) statistical parametric test. The resulting statistics are assembled into an image, that is then interpreted as a spatially extended statistical process*". In other words, at each pixel (voxel) in an image, a univariate statistical test (e.g., t-test) is applied and the resulting values of the test statistic at each pixel are then displayed as a map. The underlying spatial correlation of the map is used in creating a multivariate statistical model that describes that map and this model can be used for inference. Typically, the resulting map is analyzed using theory that underlies stationary Gaussian fields and techniques developed for excursion sets of these fields. Properties of truncated Gaussian fields (e.g., Adler and Hasofer 1976; Adler 1981; Adler and Taylor 2007; Adler et al. 2009) serve as the basis of the SPM techniques.

To date, the SPM approach has not been applied outside of medical imaging, but it appears to be a technique that could be successfully applied in a number of areas of interest in the earth and environmental sciences. The goals of this work are to both describe the basis of SPM and then apply SPM to example problems.

15.2 Anomaly Detection with Statistical Parametric Mapping

Anomaly detection is defined here as the identification of a region in time and/or space that is anomalous in its shape, size (duration) and/or values within the region (intensity). Two modes to anomaly detection in spatial-temporal data sets can be defined: (1) Anomaly detection in an online mode where prior data are used to predict future values of the measured variable and anomalies occur in areas and/or times where the predictions are inconsistent with the corresponding measurements; (2) Anomaly detection as the difference between two classes of data where differences in some treatment or external forcing condition is suspected to cause a difference in the measured variable. The anomalies in this case are significant differences in measured variables observed with and without activation of the external condition. This latter case is the focus of the work in this chapter.

Specifically, an ensemble of geologic models can be created in 1, 2 or 3 dimensions where each member of the ensemble is associated with a specific "treatment" or "result" that can be used to group ensemble members into separate classes. As examples:

- An ensemble of 3D geostatistical realizations of porosity can be created conditioned to a single set of observations where two different variogram models, both of which fit the available data equally well, are used. The different variogram models constitute a "treatment" and the question arises as to whether or not the treatments create significant differences (anomalies) in the resulting realizations and where, spatially, those differences occur.
- Petrophysical logs from different wells intersecting the same reservoir constitute an ensemble of 1-D measurements. When split into groups based on the result of which wells produced a threshold amount of petroleum and which did not, the question arises as to whether or not the petrophysical log profiles are significantly different between the groups? If they are, what portions of the log create this significant difference?

Two measures of anomaly detection can be employed: omnibus and localized (Worsley et al. 1992). Omnibus detection uses a set of calculations to determine if the current curve, map or volume, taken as a whole, is anomalous. Localized detection determines the specific location(s) within the study domain where the anomaly occurs and are the focus of this work.

Anomaly detection is not done directly on the observed generated or observed ensemble members, but on a difference between groups of members as defined through the treatment or result. Here, the differences are calculated as the differences of two average values. The averages are calculated at each point, pixel or voxel within the domain using standard univariate statistical tests (e.g., t-test). Each pixel-wise average is calculated over a set of ensemble members created under a specific condition (treatment) or generating a specific result. For example, in studies of the human brain, images are often collected under "resting" and "stimulated" conditions and the average image from each condition is then used to create a difference map.

SPM was developed to directly address the problem of spatial correlation in statistical testing. Direct application of most statistical tests requires independence of the observations, but for many problems, including those studied here, correlation between adjacent observations is the norm. Therefore, the results of the statistical tests for adjacent, or even nearby, pixels cannot be effectively evaluated using standard techniques. SPM considers a single map comprised of the results of all local (pixel-wise) statistical tests and provides several measures for comparison of the values in the map to critical threshold levels.

15.2.1 *MultiGaussian Fields*

The basis of the SPM approach is the analysis of the number, size and degree of excursions from a multiGaussian (mG) random field. For a concise, statistical description of mG fields, see Adler et al. (2009, p. 27). Stationary multiGaussian fields are fully defined by a mean and covariance matrix. In a practical sense, values at each pixel are defined with a Gaussian distribution. The correlation between those multiple distributions is defined by the covariance. Spatial correlation can be added to an uncorrelated field through the convolution of a smoothing kernel with an uncorrelated (white noise) field. As an example, the 2D Gaussian kernel is defined as

$$G(x, y) = \frac{1}{2\pi |\Sigma|^{1/2}} exp\left(-\frac{1}{2} d\Sigma^{-1} d^T \right)$$

where d is the distance vector containing distances d_x and d_y from any location (x, y) to the origin of the Gaussian function x_0, y_0 (here $(0, 0)$ for the standard normal distribution). In this work, the covariance matrix, $\Sigma = \sigma^2 I$, (I = identity matrix) is diagonal for the specific case of the kernel being aligned with the grid axes.

An often-used measure of the spatial bandwidth of a smoothing kernel in the image processing literature is the "full width at half maximum" (FWHM). For the Gaussian kernel above, the FWHM is:

$$FWHM = \sigma\sqrt{8\ln(2)}$$

If the mG field is not created, but is obtained from some type of imagery or other analyses, then there is no known underlying kernel and it is necessary to estimate the FWHM directly from the image. Estimation can be done using the covariance matrix of the partial derivatives of the image values, T, with respect to the discretization of the image. In 2D, the covariance matrix is:

$$\Lambda = \begin{bmatrix} Var\left(\frac{\partial T}{\partial x}\right) & Cov\left(\frac{\partial T}{\partial x}, \frac{\partial T}{\partial y}\right) \\ Cov\left(\frac{\partial T}{\partial x}, \frac{\partial T}{\partial y}\right) & Var\left(\frac{\partial T}{\partial y}\right) \end{bmatrix}$$

This covariance matrix can be interpreted as a measure of the roughness/smoothness of the image.

Estimation of Λ can be achieved through several approaches and here the simple relationship defined by Worsley et al. (1992) between the FWHM values in each of the principal directions and Λ is utilized. The derivatives in the covariance matrix of an image can be approximated numerically in each spatial dimension with differences between adjacent pixels are calculated as:

$$Z_{xi}(x, y) = \{T_i(x + \delta x, y) - T_i(x, y)\}/\delta_x$$
$$Z_{yi}(x, y) = \{T_i(x, y + \delta y) - T_i(x, y)\}/\delta_y$$

where δ_x and δ_y are the dimensions of the image pixels in the x and y directions. The variances and covariances of the differences are then used to approximate the variances and covariances of the derivatives:

$$V_{xx} = \sum_{i,x,y,z} Z_{xi}(x, y, z)^2 / N(n-1)$$
$$V_{yy} = \sum_{i,x,y,z} Z_{yi}(x, y, z)^2 / N(n-1)$$
$$V_{xy} = \sum_{i,x,y,z} \{Z_{xi}(x, y, z) + Z_{xi}(x, y + \delta_y, z)\}\{Z_{xi}(x, y, z) + Z_{xi}(x + \delta_x, y, z)\}/4N(n-1)$$

These variance and covariance estimates are used to estimate Λ:

$$\Lambda = \begin{bmatrix} V_{xx} & V_{xy} \\ V_{xy} & V_{yy} \end{bmatrix}$$

Finally, the FWHM in the X and Y directions are calculated as:

$$FWHM_x = \sqrt{\frac{4\ln(2)}{V_{xx}}}$$
$$FWHM_y = \sqrt{\frac{4\ln(2)}{V_{yy}}}$$

15.2.2   Calculating the SPM

The Statistical Parametric Map is the difference image between individual pairs of images or average images, which is typically transformed from a map of t-statistics to a map of Gaussian Z-score values. The different methods used in this study for calculating the SPM are described in this section.

15.2.2.1   Conditional Differences

The *t*-test and *t*-statistic are used exclusively in this chapter for the conditional differences between two ensembles and a review of the *t*-statistic is provided in the Appendix. It is noted that other statistical tests and their resulting test-statistics, e.g., χ, Z, f, as well as measures of correlation can also be used as the basis of an SPM. For the *t*-tests employed here, a location (pixel)-specific calculation of the standard deviation is used. Another approach is to calculate the pooled standard deviation across the image (image-based) and arguments for using the image-based standard deviation are given by Worsley et al. (1992). In typical applications, the number of observations under each condition is small, near a dozen, and therefore the effective degrees of freedom for $T(x, y)$ is generally small and needs to be used in the transformation of the *t*-field to a standard normal Gaussian Z-field.

The cumulative probability of a *t*-statistic is found from the *t*-distribution function with the appropriate degrees of freedom. This probability is then used with the inverse of the Gaussian distribution function to get the z-score value:

$$P(Y \leq y) = T(y; v)$$
$$z = G^{-1}(P(Y \leq y))$$

The resulting fields are now multiGaussian SPM's and the anomaly detection algorithms developed for SPM analysis can be applied.

15.2.2.2   Isolated Regions of Activation

Anomaly detection here is focused on the number, size and location of regions within an SPM that is a curve/image/volume that exceed a given threshold level, u. These regions are known as "regions of activation", "regions of exceedance" or "excursions". The numbers, sizes and locations of these excursions are then compared against a reference model of the expected expression of such regions. Truncation of a Gaussian field at a threshold u defines the u-level excursion set:

$$X_u = \{x \in R^D : Y(x) \geq u\}$$

A large body of literature on the properties of excursion sets (regions of exceedence) in Gaussian random fields is available (e.g., Adler et al. 2009; Friston et al. 1994; Lantuejoul 2002). Friston et al. (1994) characterize three related properties of excursion sets in truncated Gaussian random fields:

N the number of pixels above the truncation threshold, u,
m the number of distinct regions (inclusions) above the threshold, and
n the number of pixels in each region,

with expectation relationship $E[N] = E[m]E[n]$. For threshold value, u, the number of cells above that threshold, N, is provided by the Gaussian cdf and the size of the domain, S:

$$E[N] = S \int_u^\infty (2\pi)^{-1/2} e^{-z^2/2} dz$$

A measure of the number of isolated regions above the threshold can be obtained from the Euler Characteristic, EC. In two dimensions, the EC represents the number of connected excursion sets in the domain minus the total number of holes within those sets. Therefore, EC goes to 0.0 at $u = 0$ and EC becomes negative when $u < 0.0$ as the truncated field represents a single domain-spanning set containing a large number of holes. In 2D, and at relatively high truncation thresholds, EC is equivalent to the number of regions above the threshold, $E[m]$.

$$E[m] = EC = \left| (2\pi)^{-((D-1))/2} W^{-D} S u^{(D-1)} e^{u^2/2} \right|$$

where D is the dimension of the domain and W is an alternative measure of the spatial correlation of the mG field defined as a fraction of the FWHM:

$$W = FWHM / \sqrt{4\ln(2)}$$

For a given threshold, u. the average area of the individual regions is found from the expectation relationship:

$$E[n] = E[N]/E[m] = E[N]/[EC]$$

Figure 15.1 compares a direct calculation of EC on a multiGaussian field using the Matlab Image Processing toolbox (Matlab 2009) with estimates made using the Euler characteristic equation above across a range of u values increasing from left to right. Deviations between the calculated and estimated number of excursions indicate deviations from the definition of a multiGaussian field. The corresponding binary fields (500 $\times$ 500 cells) are also shown for several representative threshold

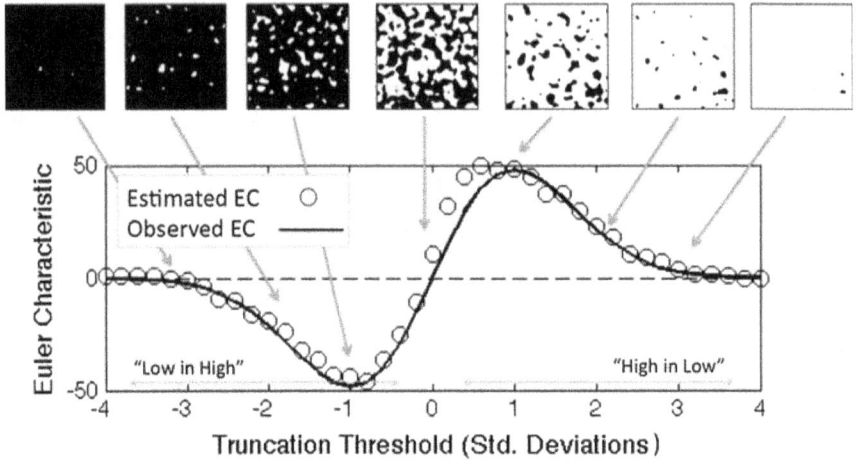

Fig. 15.1 Observed (calculated) and estimated Euler characteristic for a mG field as a function of the truncation threshold, u. The excursion sets for u > 0 are black regions in the binary fields at the top of the image (after McKenna et al. 2011)

values. Note, that typically the extreme ends of the graph corresponding to u values (truncation thresholds) with absolute values of 2.5 or greater are of interest.

15.2.3   *Localized Anomaly Detection*

Further analysis of the excursion sets is focused on the size and location of the detected anomalies. The excursion set maps themselves can be examined to determine the location of where the excursions are occurring. An extremely localized, yet very strong anomaly will be of interest. An anomaly with a much lower amplitude but greater spatial extent may also be of interest. The definition of spatial extent (size) of any anomaly is defined relative to the spatial correlation length of the field in which it is detected. The size of the anomaly is expressed through truncation of the field at a threshold value and defining the size of the excursion regions above that threshold.

In general, the significance of any anomaly in a spatial field is a function of its amplitude (intensity or strength) and its spatial extent (size). The observed SPM is compared against a specified multivariate spatial random field with a defined correlation length that serves as the model of the null hypothesis for the differences between two ensembles of spatial fields. Truncation of the observed SPM at a given threshold level creates regions of excursions above that threshold and the significance of the number and size of these excursions relative to the model of the null hypothesis is calculated. As in classical statistical hypothesis testing, the p-value defines the chance that the observed anomaly would occur under the null hypothesis. Here, the focus is on identifying the largest region of excursion for a specified threshold and calculating the chances of that anomaly occurring under the null hypothesis.

The pre-processing steps and the approach used for application of statistical parametric mapping to detection of significant excursion sets is outlined here and these steps are then applied to an example problem. The focus is on the approach used for calculation of the probability that one or more regions of activation of a certain area, or larger, could have occurred by chance under the constructed mG model. The full development of this approach for medical imaging is provided by Friston (1994) and Worsley et al. (1996). Additionally, Adler (2000) and Taylor and Adler (2003) provide further development of level crossing in random fields and the relationship to the Euler characteristic.

Steps:

(1) Create an SPM through pixel by pixel application of 1-D (pixelwise) univariate statistical tests. The test statistic values resulting from this test at every point may be distributed as χ^2, t, F, or other and can be transformed into a Gaussian Z-score to create a Gaussian SPM.

(2) Smooth the resulting SPM using a Gaussian kernel. The resulting SPM created in step 1 may be coarse and noisy. A small amount of smoothing using a Gaussian filter is enough to create a smoothed SPM.

(3) Reinflate the variance. The smoothing process in Step 2 decreases the variance of the SPM and a reinflation process is used here to transform the SPM to a unit variance (1.0) for easier interpretation of results. Here, the empirical probability distribution function of the SPM after Step 2 is fit with a Gaussian Mixture Model (GMM) having three components. A quantile-preserving transform is used to transform the empirical cumulative distribution as modeled with the GMM to a cumulative Gaussian distribution. Use of the GMM better preserves the original shape of the distribution relative to simpler transforms such as the normal-score. No translation, or recentering, of the resulting Gaussian distribution is done.

(4) Calculate the characteristics of the SPM, and choose an exceedance threshold to identify regions of exceedance.

   a. Calculate the FWHM of the smoothed and transformed SPM created in Steps 1–3. The FWHM is derived from the variances and covariances of the spatial derivatives of the SPM. The resulting FWHM values are typically 5–15 times the size of the smoothing kernel used in Step 2.

   b. Identify pixels that are above/below the $\pm$ threshold value.

   c. Employ a flood-fill algorithm to determine the sizes of the separate regions of connected pixels, or regions of exceedance and label each region for both positive and negative excursions.

(5) Apply a hypothesis test to determine the probability of a particular result having occurred under the null hypothesis of the mG model. Here, a test of the chance of obtaining the size of the of largest region of exceedance (excursion) under

the null hypothesis of a Gaussian SPM with calculated FWHM is calculated. The significance of the maximum excursion size is calculated using the methods of Friston et al (1994):

a. The three main features of an SPM are: (1) the number of pixels, N, exceeding a threshold, u. Or the number of pixels in the excursion set; (2) the number, m, of regions above the threshold—the number of connected subsets of the excursion set; (3) the number, n, of pixels in each of the m subsets. The expectations of these three features are related as: $E\{N\} = E\{n\} \bullet E\{m\}$.

b. Eq. 14 of Friston et al. (1994) gives the probability of at least one excursion region having a size $\geq k$ pixels:

$$P(n_{\max} \geq k) = \sum_{i=1}^{\infty} P(m = i) \cdot \left[1 - P(n < k)^i\right]$$

$$= 1 - e^{-E[m] \cdot P(n \geq k)}$$

$$= 1 - \exp\left(-E[m] \cdot e^{-\beta k^{2/D}}\right)$$

where $\beta = [\Gamma(D/2 + 1) \bullet E[m]/E[N]]^{2/D}$ and D is the dimension of the domain.

Calculations of $P(n_{max} \geq k)$ within the (k, u) parameter space for spatial fields with two different correlation lengths (FWHM) are shown in Fig. 15.2. The role of the correlation length of the null hypothesis model is clear from Fig. 15.2 where the probability of an excursion region of 60 pixels or more is approximately 0.001 for a field with a FWHM of 9.0, but is essentially zero ($\sim 10 \times 10^{-12}$) for a field with a FWHM of 3.0.

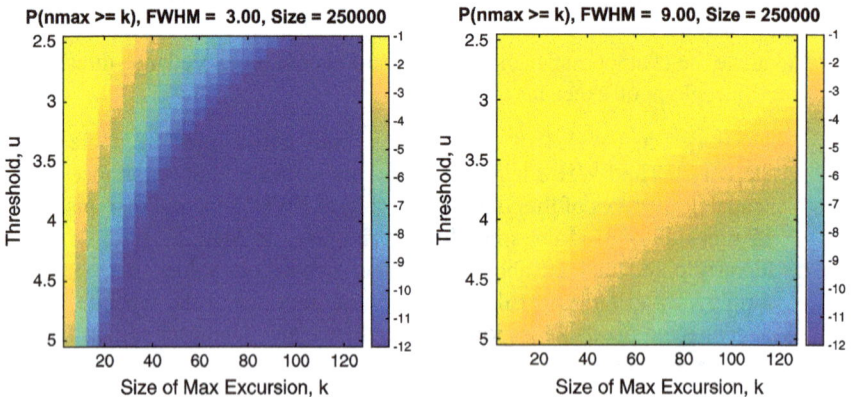

Fig. 15.2 $P(n_{\max} \geq k)$ as a function of size of the excursion region, k, and the truncation threshold, u, for fields of size 500 × 500 with an isotropic FWHM of 3.0 pixels (left) and 9.0 pixels (right). The color scale is log10($P(n_{\max} \geq k)$)

15.3   Example Problems

Two example problems are used here to demonstrate the calculations and application of SPM to detecting anomalies in spatial random fields. Both example problems are two-dimensional, but the same approaches are applicable to anomaly detection in 1-D and 3-D domains.

15.3.1   Anomaly Detection in Images

A simple simulation study designed to mimic the detection of anomalous regions in either remote sensing or geophysical imagery is used here to test a few of the SPM calculations. The focus is on identifying the largest anomaly above a specified threshold and the significance of that anomaly.

A multiGaussian field is created through geostatistical simulation. The field is comprised of square, 5 × 5 m pixels, and has an isotropic Gaussian variogram with a range of 150 pixels. The field is created in standard normal space, $N(0, 1)$ and the simulated values serve as the observed image. Measurement noise is added to the image by considering the simulated realization value, $z(x)$ to be the mean value of a local Gaussian distribution at every pixel. The standard deviation of the Gaussian at every pixel, $\sigma_z(x)$, is set to 2.0 and a Gaussian random deviate is drawn and added to $z(x)$ to create the final image. This measurement noise is added independently at every pixel (i.i.d.) and then smoothed prior to adding to the observed image. The amount of spatial smoothing of the noise term is varied and the impact on anomaly detection is examined.

Anomalies are added to the observed image within a circular region having a radius of 90 pixels and centered at the center of the image. Background values within the anomaly region are multiplied by 1.5 creating stronger negative and positive values within the region depending on the sign of the original observed values. The area of the anomaly region is 5027 pixels.

Figure 15.3 shows background images (left column) at two levels of noise smoothing and the background images with the anomalies added (right column). As would occur in any image capture process, the noise values added to each image are drawn randomly and independently from any other image prior to smoothing. This creates subtle differences between the images in each row of Fig. 15.3 even without the addition of anomalies. Detection of the presence of the anomalies through visual comparison of the left and right images in each row of Fig. 15.3 is not obvious, even when the location of the anomaly is known.

The SPM's are calculated through a pixelwise t-test for comparing two means (Appendix) between the image with and without the anomalies. These t-statistic maps are transformed to Gaussian Z maps that are the SPM (Fig. 15.4). The large anomalies in the center of the image are readily seen along with the dramatic changes in the results due to the increased spatial correlation of the noise

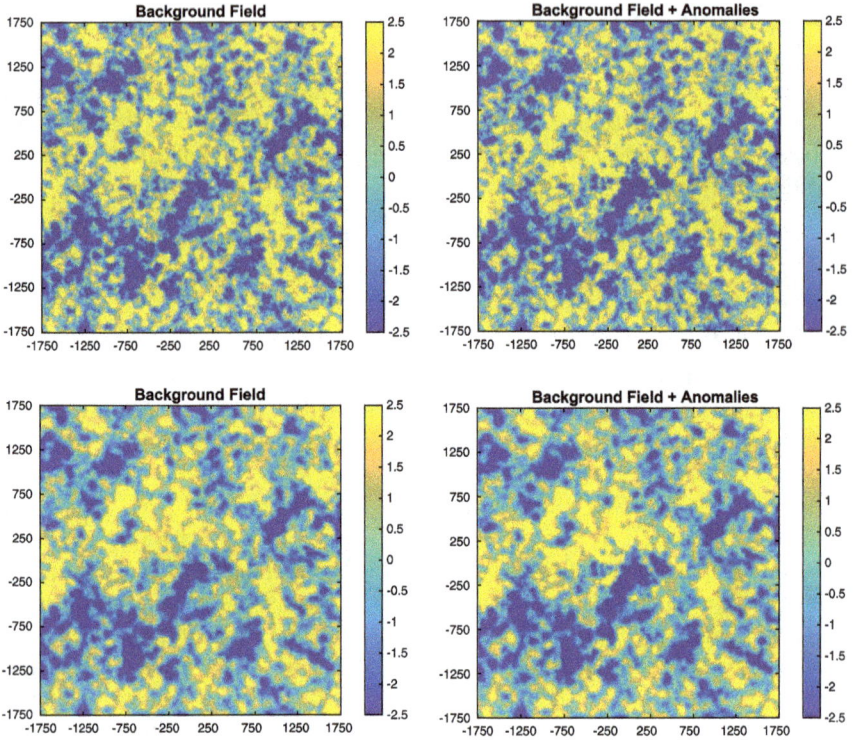

Fig. 15.3 Background fields without (left column) and with (right column) added anomalies with a smoothing kernel size $\sigma = 1.5$ pixels (top row) and $\sigma = 7.5$ pixels (bottom row). Color scale units are arbitrary in this example

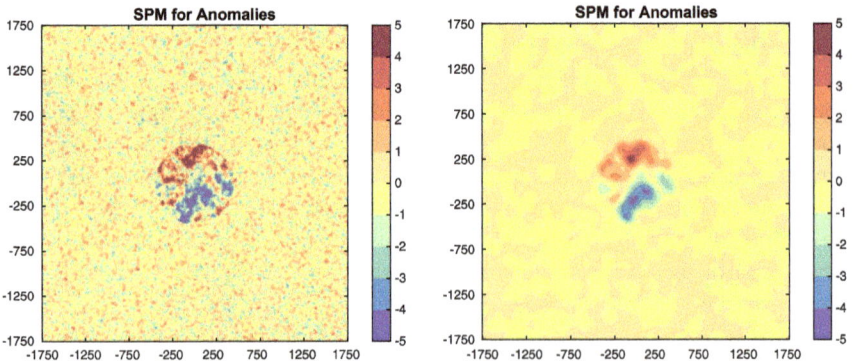

Fig. 15.4 SPM's for the case of smoothing with a filter bandwidth of $\sigma = 1.5$ (left) and $\sigma = 7.5$ pixels (right). The color scale is in standard deviations away from the mean of zero

Table 15.1 Results of SPM analysis for four levels of noise smoothing

Bandwidth of smoothing filter (pixels)	1.5	3.5	5.5	7.5
FWHM	5.0	23.5	45.0	71.6
Size Max Positive (pixels)	2061	2886	1878	2075
$P(n_{max} > k)$ Positive	$<1.0 \times 10^{-16}$	1.3×10^{-06}	0.237	0.405
Size Max Negative (pixels)	3640	3664	3772	3417
$P(n_{max} > k)$ Negative	$<1.0 \times 10^{-16}$	1.5×10^{-08}	0.015	0.200
# regions >+ threshold	238	5	5	1
# regions <− threshold	212	5	2	1
Max SPM value (standard deviations)	8.28	6.45	5.25	4.55
Min SPM value (standard deviations)	−7.83	−6.23	−5.43	−4.58

component with increased smoothing. Additional SPM's are created at intermediate levels of smoothing but are not shown here. Results from all levels of noise smoothing are shown in Table 15.1.

A threshold of ±2.5 standard deviations is applied to the SPM's and the excursion regions for the two extreme levels of noise smoothing are shown in Fig. 15.5. There are over 200 positive and 200 negative excursions for the smallest amount of noise smoothing and only 1 positive and 1 negative excursion at the largest amount of smoothing. The size of the excursions that are due to the added anomalies clearly stands out in the left image of Fig. 15.5. Table 15.1 also shows how the maximum and minimum images in the SPM decrease with increased levels of noise smoothing.

With increased smoothing of the noise, the FWHM of the image increases from 5.0 to ~ 72 pixels (Table 15.1). While the size of the largest positive and negative excursions remains approximately constant near 2000 and 3600 pixels, respectively, the p-value for excursions of that size occurring in the image changes dramatically. At the lowest level of smoothing, the chances of getting excursions of size 2061 or 3640 pixels under the Gaussian random field model with a FWHM of 5.0 are essentially zero ($< 1.0 \times 10^{-16}$). However, getting excursion regions of a similar size occurring under greater smoothing of the noise and a FWHM of 71.6 pixels is relatively common at 40 and 20%, respectively. These results demonstrate the strong dependence of $P(n_{max} >= k)$ on the spatial correlation of the field.

15.3.2   Ground Water Pumping

A general problem in a number of geoscience disciplines is the case where an ensemble of inputs is used in a calculation to provide a probabilistic result to a particular question. The calculation can be relatively simple or complex, but acts as a transfer function to transfer uncertainty in spatially distributed physical properties to uncertainty in an outcome of interest. Examples include groundwater models

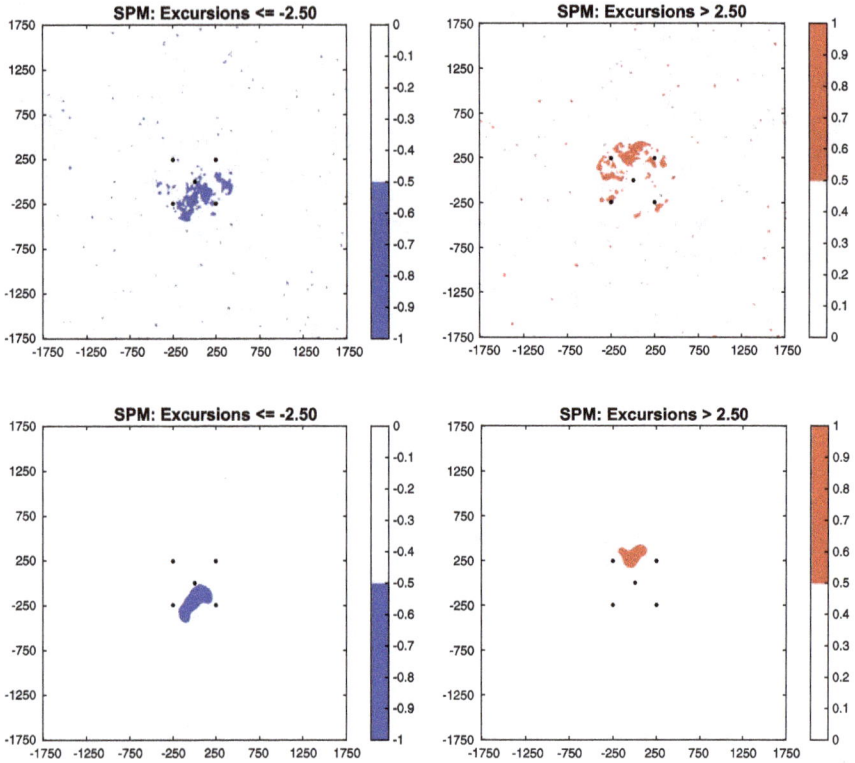

Fig. 15.5 Regions of excursion below a threshold of −2.5 (left column) and above 2.5 (right column) for images with noise smoothed using a filter of σ = 1.5 (top row) and σ = 7.5 (bottom row)

transferring uncertainty in hydraulic conductivity and recharge into radionuclide transport times; reservoir simulators transferring uncertainty in permeability and porosity into estimated recoverable oil; and simple spatial integration to transfer uncertainty in soil nutrient levels into estimating total crop yield for an agricultural field.

Here, a ground water example problem is used with the SPM approach to detect significant differences between two groups of an ensemble of spatial random fields of transmissivity. The ensemble is split into groups that create high results and all others. The SPM approach is used here to identify statistically significant features within the ensemble of input fields responsible for the specific results. This approach can be considered identification of the significant features in the random fields responsible for a specific result of a process that integrates across the entire field.

15.3.2.1   Problem Setup

The ground water problem is motivated by the regulatory issue of impacts on a nearby wetland due to pumping from a planned water supply well. Well test criteria dictate that the pressure drop (drawdown) at a location 353 m to the northwest of the pumping well must be <2.00 m after pumping at a rate of 250 m³/h for 48 h. To simulate the aquifer test, a 12 × 12 km square domain, with zero-flux boundaries on the north and south and constant-head boundaries on the east and west is defined. Prior to pumping, the fixed head boundaries create steady state flow across the domain. A constant transmissivity, T, of 10.0 m²/h is assumed across the majority of the domain. This constant value is replaced by a heterogeneous T field within the center of the domain. The heterogeneous field is 3500 × 3500 m with 5 × 5 m cells. A large pumping well is set in the center of the domain.

The aquifer is confined in this area and the mean and spatial co-variance of the transmissivity can be estimated from other studies in aquifers of similar age and depositional history. The log10 values of transmissivity within the heterogeneous domain are simulated as a multiGaussian field with an isotropic Gaussian variogram with range 250 m and nugget of 5% of the sill. Transmissivity at the well location is considered known and provides the only conditioning point within the domain. A total of 200 realizations are created, and the 2D, confined, transient ground water flow equation is solved using finite differences on each realization:

$$\frac{\partial h(x,y)}{\partial t} = \frac{1}{S(x,y)} \cdot \left(\frac{\partial}{\partial x} T(x,y) \frac{\partial h}{\partial x} \right) + \left(\frac{\partial}{\partial y} T(x,y) \frac{\partial h}{\partial y} \right) \pm Q(x,y)$$

where (x, y) indicates the spatial location, h (L) is the head (pressure), t is time and Q (L^3/T) are sources or sinks—here the pumping rate at the well. Transmissivity, T (L^2/T), is spatially heterogeneous within the central domain and for the calculations here, storativity, S (−) is set to a single value of 1.0×10^{-05} across all locations in the aquifer. The initial conditions for the transient simulation are taken from a steady state head solution using the same input T field. Three example transmissivity realizations and maps of the resulting drawdowns after 48 h of pumping are shown in Fig. 15.6. Figure 15.6 demonstrates that the heterogeneous T field strongly impacts the resulting pressure response in a non-linear manner.

15.3.2.2   Results

For each ground water simulation, the drawdown at the test location (353 m NW of the pumping well) at 48 h is recorded and compared to the regulatory limit, R, of 2.00 m. The T realization is placed into one of two classes: those that meet the pressure drop limit, drawdown <= R, and those that exceed the limit. After 200 ground water simulations, the pixelwise mean and standard deviation within each

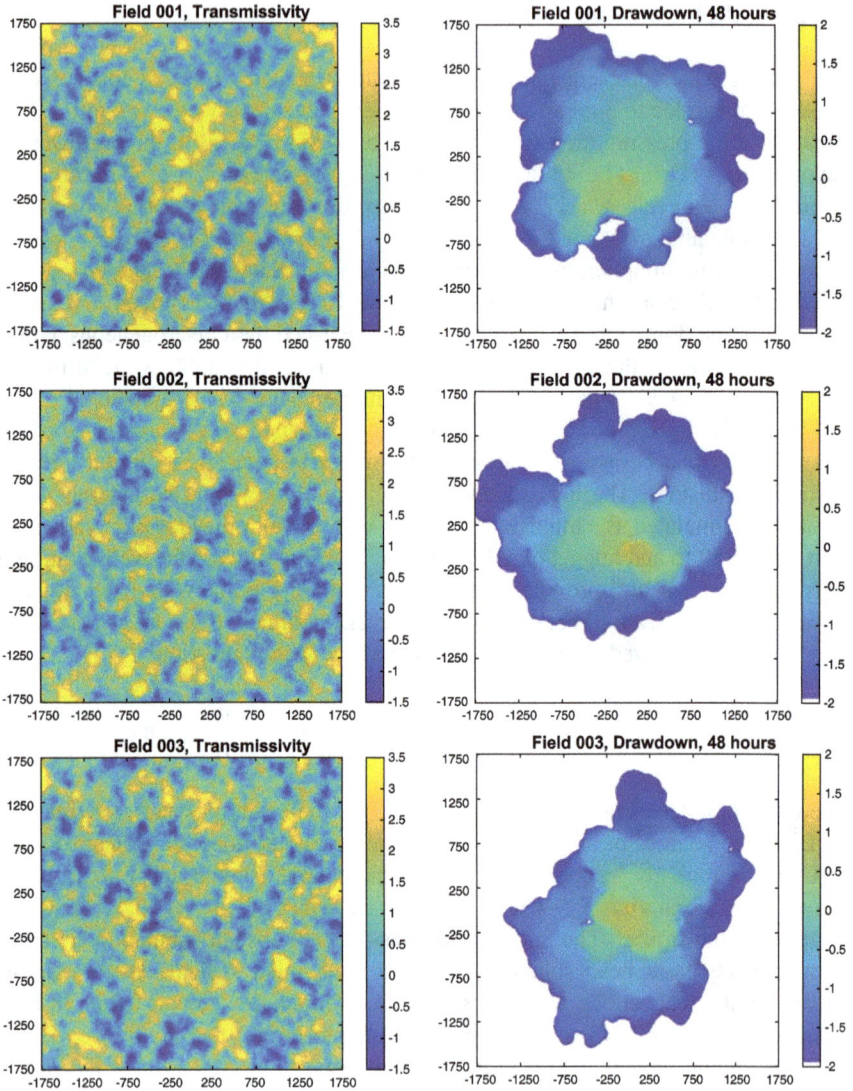

Fig. 15.6 Three example transmissivity fields (left column) and the corresponding ground water drawdown levels after 48 h of pumping (right column). The color scales define log10 T in m^2/h and log10 drawdown in meters

class are calculated (Fig. 15.7). These four maps provide the input to a two-sample t-test to determine the difference between two means. The resulting map of t-statistics is the SPM. Here the t-statistics are smoothed with Gaussian kernel and transformed to Z-statistics and the Z-score SPM is shown in Fig. 15.8.

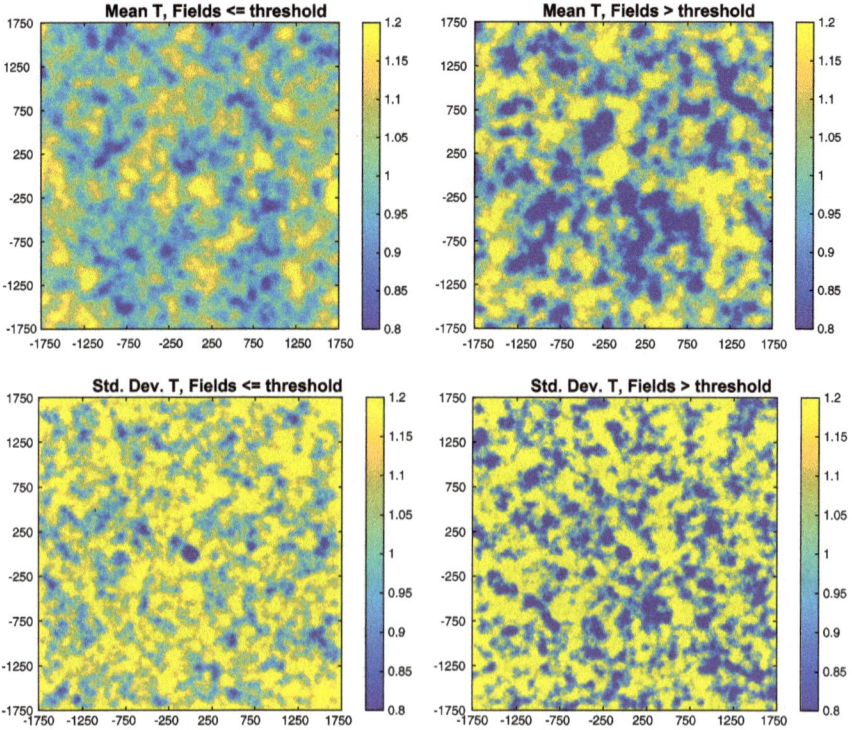

Fig. 15.7 The mean (top row) and standard deviation (bottom row) fields for the transmissivity realizations that create drawdown <=2.00 m (left column) and drawdown >2.00 m (right column). The color scales show $\log10(T)$ in m^2/h

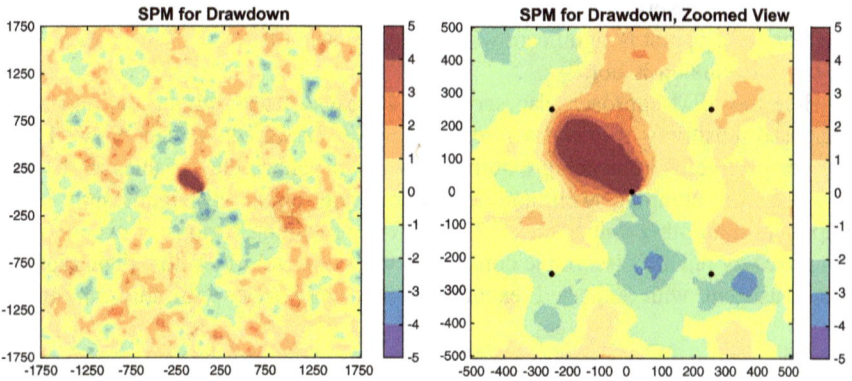

Fig. 15.8 SPM for the difference between realizations. Full field is shown on the left and a zoomed in view of the central field on the right. Color scale is in standard deviations away from the mean of zero

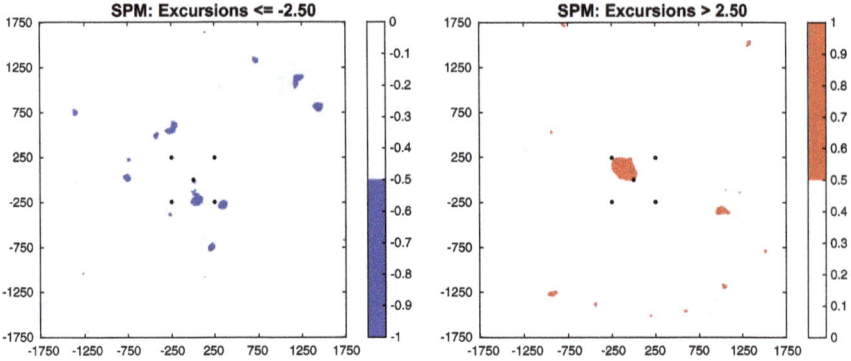

Fig. 15.9 Regions of excursion below a threshold of −2.5 (left) and above 2.5 (right)

The SPM is calculated at every pixel as the mean T value of the fields that created drawdowns exceeding the regulatory threshold, R, minus those that resulted in drawdowns less than or equal to the threshold: $T_{>R} - T_{\leq R}$. This convention creates a positive value in the SPM in an area where higher T values are associated with realizations that created exceedance of R and negative values where higher T values created drawdowns $\leq R$. Figure 15.8 shows regions of positive and negative values, but the dominant anomaly is a high SPM value between the pumping well and the observation point to the northwest. For this example, 155 realizations (77.5%) created drawdowns $\leq R$ and 45 (22.5%) created drawdowns that exceeded R.

The SPM is truncated at a threshold of $\pm 2.5\sigma$ and the excursion regions are defined (Fig. 15.9). The size of the largest excursions and the probability of them occurring under the mG model are shown in Table 15.2. The SPM has a FWHM of 111.5 m (22.3 pixels). The large positive excursion between the pumping well and the monitoring point is significant with a p-value near 1.0×10^{-04} while the largest negative excursion is not.

Here the SPM approach also serves as a means of determining the regions of increased sensitivity of drawdown to the T values. As expected, when viewed from the perspective of influencing extreme drawdown values, the T values in the area between the pumping well and the monitoring point are significantly more important than other values in the T field. The remaining regions of excursion do not have any readily discernible connection to the ground water flow dynamics and are consistent with expected excursions in a mG field with this amount of

Table 15.2 Largest positive and negative excursions in ground water example SPM

FWHM (m)	111.5	
Excursions	Positive	Negative
Largest excursion (pixels)	2369	713
$P(n_{max} > k)$	$\sim 1.0E^{-04}$	0.198

correlation. In practice, the large positive excursion region in the SPM can be used to focus resources for additional data collection, e.g., geophysical survey and/or additional wells.

15.4   Summary

There is a large amount of work reported in the functional MRI literature on the detection of anomalies in spatially correlated fields using SPM. Apart from some work in astrophysics, this SPM work has generally been restricted to medical imaging. The body of knowledge around SPM and the statistical approaches developed for fMRI can be readily applied to problems in the earth and environmental sciences. This chapter reviews some of the major developments from the fMRI literature and demonstrates their application with an image anomaly detection problem and a ground water modelling problem. A strong advantage of SPM is that it directly addresses the challenge of enabling hypothesis testing, including calculation of the significance of the results, in spatially correlated fields.

The example problems chosen here emphasized defining the significance of the largest, positive and negative, anomaly in each SPM. The SPM framework also supports hypothesis testing on non-localized, "omnibus", features such as the maximum/minimum value of the SPM, the number pixels exceeding the threshold and the number of excursion regions within the SPM. Additionally, hypothesis testing of localized, "focal", features is also supported including hypothesis testing of the occurrence of any size excursion.

The example problems used here relied on the underlying images being realizations of mG fields, but that is not a requirement. It is the map of the test statistic values defining the differences between fields that is modelled as a mG field, and that flexibility makes SPM applicable to a very general set of problems as the mG model is a standard for differences between images. For example the same approach could be used to compare geologic models with discrete features. Future work will consider the application of other statistical tests within the SPM framework.

Appendix: Conditional Differences

The t-test is a traditional measure of the difference between two means (e.g., Walpole and Myers 1989). Quite simply, the t-statistic is the difference between two values, at least one of which is a population or sample mean, normalized by the standard error of the mean:

$$t = \frac{\overline{X} - \mu}{s_e} = \frac{\overline{X} - \mu}{s\sqrt{1/n}}$$

where $\overline{X}$ is a sample mean, μ is a population mean, s_e is the standard error of the mean which is the standard deviation of the observations, s, that make up the data vector X multiplied by the square root of 1 over the number of samples within X. The cumulative probabilities for any value of t are available from the Student's t distribution and require knowledge of the degrees of freedom, ν, within the test. For the analyses done here, ν is generally $n - 1$.

In the case of comparing two sample means to each other at each location, i.e., A (x, y) and $B(x, y)$, instead of comparing a sample mean to a theoretical population mean, the value of s_e must be calculated from both sample sets as:

$$S_e = S_p \sqrt{\frac{1}{n_1} + \frac{1}{n_2}}$$

where n_1 and n_2 are the number of images that were used in calculating the average maps A and B and s_p is the average pooled standard deviation:

$$S_p(x, y) = \sqrt{\frac{(n_1 - 1)s_1^2(x, y) + (n_2 - 1)s_2^2(x, y)}{n_1 + n_2 - 2}}$$

Here we are assuming that n_1 and n_2 are constant for all locations and therefore not a function of (x, y). The t-statistic image (map), based on the pooled standard deviation, is:

$$t(x, y) = \frac{\Delta(x, y)}{S_p(x, y)\sqrt{\frac{1}{n_1} + \frac{1}{n_2}}}$$

References

Adler RJ, Hasofer AM (1976) Level crossings for random fields. Ann Probab 4:1–12
Adler RJ (1981) The geometry of random fields. Wiley, New York
Adler RJ (2000) On excursion sets, tube formulas and maxima of random fields. Ann Probab 10(1):1–74
Adler RJ, Taylor JE (2007) Random fields and geometry. Springer
Adler RJ, Taylor JE, Worsley KJ (2009) Applications of random fields and geometry: foundations and case studies. http://webee.technion.ac.il/people/adler/publications.html. Accessed Jan 2009
Anselin L (1995) Local indicators of spatial association—LISA. Geogr Anal 27(2):94–115
Brett M, Penny W, Kiebel S (2003) An introduction to random field theory, chapter 14. In: Frackowiak RSJ, Friston KJ, Frith C, Dolan R, Friston KJ, Price CJ, Zeki S, Ashburner J, Penny WD (eds) Human brain function, 2nd edn. Academic Press

Byers S, Raftery AE (1998) Nearest-neighbor clutter removal for estimating features in spatial point processes. J Am Stat Assoc 93:577–584

Cressie N, Collins LB (2001) Patterns in spatial point locations: local indicators of spatial association in a minefield with clutter. Nav Res Logist. https://doi.org/10.1002/nav.1022

Friston KJ, Worsley KJ, Frackowiak RSJ, Mazziotta JC, Evans AC (1994) Assessing the significance of focal activations using their spatial extent. Hum Brain Mapp 1:210–220

Friston KJ, Holmes AP, Worsley KJ, Poline J-P, Firth CD, Frackowiak RSJ (1995) Statistical parametric maps in functional imaging: a general linear approach. Hum Brain Mapp 2:189–210

Gilbert RO (1987) Statistical methods for environmental pollution monitoring. Van Nostrand Reinhold, New York, p 320

Goovaerts P, Jacquez GM, Marcus A (2005) Geostatistical and local cluster analysis of high resolution hyperspectral imagery for detection of anomalies. Remote Sens Environ 95(3): 351–367

Goovaerts P (2009) Medical geography: a promising field of application for geostatistics. Math Geosci 41:243

Lantuejoul C (2002) Geostatistical simulation: models and algorithms. Springer, Berlin, p 256

Matlab (2009) Image processing toolbox. Mathworks Inc, Natick, MA, USA

McKenna SA, Ray J, van Bloemen Waanders B, Marzouk Y (2011) Truncated multigaussian fields and effective conductance of binary media. Adv Water Resour 34:617–626

Stein DWJ, Beaven SG, Hoff LE, Winter EM, Schaum AP, Stocker AD (2002) Anomaly detection from hyperspectral imagery. IEEE Signal Process Mag 19(1):58–69

Taylor JE, Adler RJ (2003) Euler characteristics for Gaussian fields on manifolds. Ann Probab 31(2):533–563

Walpole RE, Myers RH (1989) Probability and statistics for engineers and scientists, 4th edn. MacMillan Publishing Co., New York, p 765

Worsley KJ, Evans AC, Marrett S, Neelin P (1992) A three-dimensional statistical analysis for rCBF activation studies in human brain. J Cereb Blood Flow Metab 12:900–918

Worsley KJ, Marrett S, Neelin P, Vandal AC, Friston KJ, Evans AC (1996) A unified statistical approach for determining significant signals in images of cerebral activation. Hum Brain Mapp 4:58–73

16

Role of Compositional Data Analysis in Water Chemistry

Antonella Buccianti

Abstract John Aitchison died in December 2016 leaving behind an important inheritance: to continue to explore the fascinating world of compositional data. However, notwithstanding the progress that we have made in this field of investigation and the diffusion of the CoDA theory in different researches, a lot of work has still to be done, particularly in geochemistry. In fact most of the papers published in international journals that manage compositional data ignore their nature and their consequent peculiar statistical properties. On the other hand, when CoDA principles are applied, several efforts are often made to continue to consider the log-ratio transformed variables, for example the centered log-ratio ones, as the original ones, demonstrating a sort of resistance to thinking in relative terms. This appears to be a very strange behavior since geochemists are used to ratios and their analysis is the base of the experimental calibration when standards are evolved to set the instruments. In this chapter some challenges are presented by exploring water chemistry data with the aim to invite people to capture the essence of thinking in a relative and multivariate way since this is the path to obtain a description of natural processes as complete as possible.

16.1 Water Chemistry Data as Compositional Data

When geochemical data are analysed by using statistical methods, several units can be used to express concentrations and a first discussion of their compositional nature is reported in Buccianti and Pawlowsky-Glahn (2005). The usual units of measurement include milligrams per liter (mg/L), parts per million by weight (ppm), parts per billion by weight (ppb), millimole per liter (mmol/L), and

A. Buccianti (✉)
Department of Earth Sciences, University of Florence, Via G. La Pira 4,
50121 Florence, Italy
e-mail: antonella.buccianti@unifi.it

A. Buccianti
CNR-IGG, Unit of Florence, Florence, Italy

milliequivalent per liter (meq/L). The ppm and mg/L units are numerically equal if the density of the water sample is 1 g/cm^3, as in pure water. Samples can be converted from mg/L to ppm by multiplying each component by the density of water. The term mmol/L indicates the number of ions or molecules in the water when multiplied by Avogadro's number (the number of molecules in a mole of material, 6.023×10^{23}). The measure mg/L is converted to mmol/L by dividing by the atomic or molecular weight. To express concentration by meq/L (electrical charges are considered), mmol/L is multiplied by the charge of the ions. In each case the base of the calculus is given by the content of some chemical species referred to a given weight or volume then multiplied by a constant (atomic or molecular weight, electrical charges).

These types of data describe parts of some whole and even if proportions are expressed as real numbers, they cannot be interpreted, or even analysed, as real data. It is well known that this practice can lead to paradoxes and/or misinterpretations (e.g. intervals covering negative proportions, spurious correlations) already discussed a century ago (Pearson 1897), but mostly forgotten and neglected over the years (Chayes 1960).

No other ways are possible to compare different samples from dissimilar sites and times, as is usually required. Thus the compositional nature of the experimental data is an intrinsic property related to their origin (e.g. instrument calibration) and to the necessity of making comparisons to investigate the genesis of environmental variability. As directional (circular) observations (Fisher 1995) compositional data move in a constrained sample space called *simplex* (Aitchison 1986):

$$S^D = \{x = [x_1, x_2, \ldots, x_D] | x_i\}, > 0, \ i = 1, 2, \ldots, D; \ \sum_{i=1}^{D} x_i = \kappa \qquad (16.1)$$

where the D components of the vector S^D are called parts (variables) of the composition. The value of κ depends of the units of the measurement or rescaling procedure, and usual values are 1 (proportions), 100 (%), 10^6 (ppm) or similar. Note that it is not necessary to have $\sum_{i=1}^{D} x_i = \kappa$ (closed data) to obtain compositional observations. In fact, a (row) vector $x = [x_1, x_2, \ldots, x_D]$ is a D-part composition when all its components are strictly positive real numbers and carry only *relative information*. This means that the message about what is occurring is mainly contained in the ratios between the parts since the numerical value of each variable by itself is not relevant. A recent thorough analysis of the "compositional problem" can be found in Pawlowsky-Glahn and Buccianti (2011) and Pawlowsky-Glahn et al. (2015). On the other hand interesting applications on water chemistry can be found in literature (e.g. Engle and Rowan 2013, 2014; Engle and Blondes 2014; Buccianti and Zuo 2016; Owen et al. 2016; Buccianti et al. 2018; Shelton et al. 2018) where the different potentialities of the family of the log-ratio transformations are differently exploited posing at the central point of the analysis the relativity of the values and the multivariate vision. The cited papers are not exhaustive but have been chosen since they successfully focus on the use of the isometric log-ratio transformation as a way to describe the dynamics of geochemical processes.

16.2   Isometric-Log Ratio Transformation: Is This the Key to Decipher the Dynamics of Geochemical Systems?

16.2.1   Coordinates as Balances

Water present below the land surface and running above it tells the history of the environment with which it has been in contact. Rainfall and snowmelt interact with the rock of the Earth surface and percolate through the soil zone where chemical reactions with gases, minerals and organic compounds take place. Chemical reactions occur because the composition of the water is not in equilibrium with the solid phases or the gaseous component (Kleidon 2010). Thus disequilibrium drives the reactions and solutes in the water are derived from the dissolution or leaching of the solid phases and from the dissolution of gases from the air or from the oxidation of organic matter. Most of the natural systems are open and according with Nicolis and Prigogine (1989) they are characterized by dissipative structures and presence of irreversible processes. Dissipative structures contain subsystems, which permanently fluctuate until the fluctuation becomes so strong that it breaks the original system to generate a new condition, more complex and characterized by a higher level of order. The dynamics of systems being far from equilibrium requires a continuous self-organization and to maintain this condition the energy flux from the environment is higher than required for the initial state and irreversible processes can be a source of order rather than chaos. Most of the geological systems are open and dynamic, characterized by a great number of components and develop in a nonlinear way far from equilibrium (Shvartsev 2009). Particularly interesting from this point of view is the water-rock system where also synergetic properties can be found, with respect to the thermodynamical equilibrium where elements (molecules) behave independently of one another (Shvartsev 2013).

The use of the isometric log-ratio coordinates (Egozcue et al. 2003) not only allows us to manage compositional data with classical statistical tools, but also could offer a powerful tool to probe the level of self-organization of a geochemical system as a whole. When coordinates are obtained by using the sequential binary partition method (Egozcue and Pawlowsky-Glahn 2005), guided by a geochemical criterion, the analysis of their frequency distribution may represent an interesting way to understand the laws governing randomness and variability. By taking into account this consideration, an improvement of the *balance dendrogram* (Pawlowsky-Glahn and Egozcue 2001) is here presented with the aim to investigate the behavior of aqueous systems.

The sample space of D-part compositional data, the simplex, being a subset of the real space R^D, has a real Euclidean vector space structure (Billheimer et al. 2001; Pawlowsky-Glahn and Egozcue 2001; Buccianti and Magli 2011). This situation allows the representation of data in coordinates with respect to an orthonormal basis, for example following the Gram-Schmidt orthonormalization process or a Singular Value Decomposition (Egozcue et al. 2003). Since these methods often reveal coordinates not easy to interpret, *balances*, a specific type of orthonormal coordinates associated with groups of parts, have been proposed (Egozcue and Pawlowsky-Glahn 2005). This method is based on a sequential binary partition of a *D*-part composition into non-overlapping groups and when the procedure is geochemically guided it leads to coordinates easy to interpret. Moreover, it allows understanding of how the total variance is decomposed into marginal variances, thus pointing out the relationship between intra-group and inter-group compositional parts variability. For the *i-th* order of partition, the balance is

$$b_i = \sqrt{\frac{r_i \cdot s_i}{r_i + s_i}} log \frac{\left(\prod_{x_j \in G_{i1}} x_j\right)^{1/r_i}}{\left(\prod_{x_l \in G_{i2}} x_l\right)^{1/s_i}} \tag{16.2}$$

where r_i and s_i are the number of parts in the groups of numerator (G_{i1}) and denominator (G_{i2}), respectively. As we can see, the balance is defined as the natural logarithm of the ratio of geometric means of the parts in each group, normalized by the coefficient needed to obtain unit length of the vectors of the basis.

16.2.2 Behavior of Self-organizing Systems and CoDA Phylosophy

A general characteristic of self-organizing systems is robustness and resilience (Dakos et al. 2014; Dai et al. 2015). This means that they are relatively insensitive to perturbations or errors, and can show a strong capacity to restore themselves after changes (Scheffer et al. 2009, 2012). One reason for this fault-tolerance is the redundant, distributed organization so that the non-damaged regions can usually make up for the damaged ones. Within certain limits, another reason for the intrinsic robustness is that self-organization is facilitated by randomness, fluctuations or "noise" while the stabilizing effect of feedback loops guarantee resilience. The presence of feedback mechanisms generates systems that can be responsible for their own maintenance, and thus largely independent from the environment. Although in general there will still be exchange of matter and energy between systems and surroundings, the organization is determined purely internally. Thus the system is thermodynamically open, but organizationally closed. Organizational closure turns a collection of interacting elements into an individual, coherent whole. This whole has properties that arise out of its organization that can be described by the probability laws that govern the relative behaviour of its elements (van Rooij 2013). From this point of view CoDA theory appears to capture the philosophy of

this condition and the analysis of the shape of the frequency distribution of isometric coordinates should be the adequate tool (Allegre and Lewin 1995; Seely
et al. 2012; Holden and Rajaraman 2012; Buccianti and Zuo 2016).

As reported in Scheffer et al. (2012) the probability density distribution of some
variables describing the state of a system can be used to estimate how the potential
landscape is reflecting its stability properties. The shape of the probability density
function indicates where the data are more aggregated and which laws are governing the variability, giving us fundamental information about the genesis of
randomness (Agterberg 2014). In our case it will be the shape of the frequency
distribution of isometric log-ratio coordinates representing some geochemical
process that will inform us about dynamic properties of the system. In Fig. 16.1
some examples of a non-equilibrium dynamics are reported (Scheffer et al. 2009).
Conditions represented in (a) are far from a bifurcation point. The pothole in the
potential line corresponds to an area where data tend to aggregate in the density
probability distribution function. Here resilience is large since the basin of attraction
is wide and the rate of recovery from perturbations is relatively high. If the system
is stochastically forced, the resulting dynamics will be characterised by low correlation between states at subsequent time intervals. In (b) the system is closer to the
transition point and resilience decreases due to the shrinking of the attraction basin
and the low rate of recovery from small perturbations. Here the slight depression
could be related to presence of bimodality indicating presence of alternative states.
In this case the system in a stochastic environment will have a long memory for
perturbations and its dynamics will be governed by high variance and stronger
correlations between subsequent states.

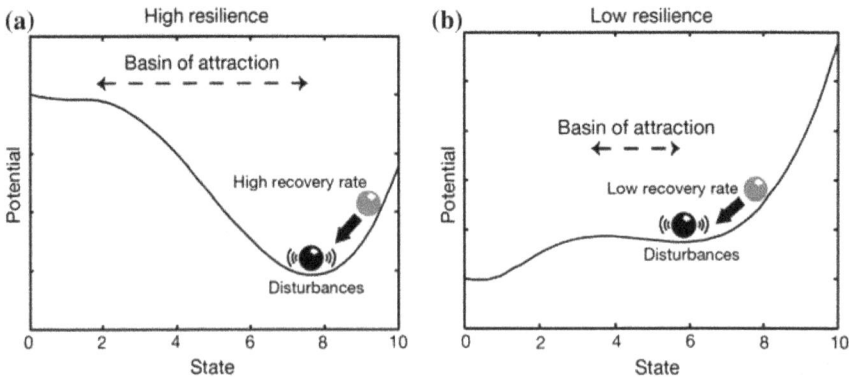

Fig. 16.1 Example of non-equilibrium dynamics (from Scheffer et al. 2009, modified). The
pothole in the potential line of diagram **a** corresponds to an area where data tend to aggregate in the
density probability distribution function. The slight depression in **b** could be related to presence of
bimodality indicating presence of alternative states (Scheffer et al. 2012)

16.3   Improving CoDA-Dendrogram: Checking for Variability, Resilience and Stability

The chemical composition of groundwaters from the Arezzo basin aquifer (Tuscany, central Italy) was analysed, as an application example, to obtain information about the dynamics of the aqueous geochemical system. The Arezzo Basin (Fig. 16.2), formed since Upper Pliocene, is a structural depression bordered to the North and to the East by the Pratomagno and Chianti belts, respectively, and to the South and to the East by two tectonic lineaments (Val d'Arbia-Val Marecchia transversal and Chitignano normal faults). Along these tectonic discontinuities CO_2-rich manifestations either seep out or are exploited by private companies down to the depth of 1000 m. Three main aquifers are recognized: (i) a relatively deep aquifer hosted in Tertiary sandstone formations; (ii) an intermediate aquifer hosted in Quaternary fluvio-lacustrine sediments and (iii) a shallow aquifer in recent alluvial sediments. The available geochemical data-base consists of about 500 samples that were collected in different dry and rainy seasons in recent years from 80 wells diffused in all the basin area. Depth of the sampling is, unfortunately, not always known and few differences can be related to seasonal changes. Physical parameters (temperature and electrical conductivity), major, minor and trace dissolved species (pH, Ca, Mg, Na, K, NH_4, HCO_3, SO_4, NO_3, NO^2, Cl, Br, F and heavy metals), oxygen and hydrogen isotopes in the water molecules and dissolved gases (including ^{13}C-CO_2) were analyzed. On the basis of Total Dissolved Solids (TDS) the waters from Arezzo aquifer can be considered mainly oligomineral and medium-mineral, whereas mineral waters are almost exclusively associated with

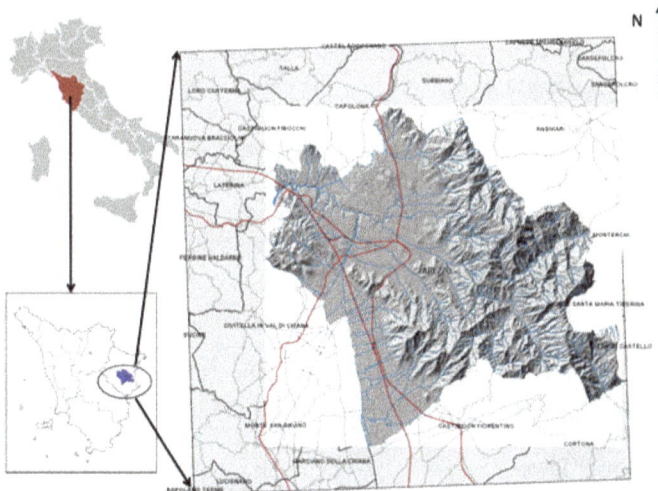

Fig. 16.2 The hydrographic system of the Arezzo basin (Tuscany, central Italy) (http://sit.comune. arezzo.it/normativa/index.php?normativa=_ps&mappa=ps_b11a)

CO_2-rich wells. From a classification point of view, Ca(Mg)-HCO_3 is by far the most representative geochemical facies, followed by Na(K)-HCO_3, Ca(Mg)-SO_4 and Na(K)-Cl types. It is noteworthy to point out here that the Na(K)-HCO_3 waters, whose origin is related to the presence of CO_2-rich waters that favor cation exchange processes with clay minerals contained in the sedimentary formations, are aligned along the Val d'Arbia-Val Marecchia transversal tectonic system.

In Table 16.1 the sequential binary partition process to construct the isometric log-ratio coordinates is reported. The first coordinate could represent the balance between the most important chemical reactions involving carbonatic and silicatic rocks (Ca^{2+}, Mg^{2+}, Na^+, K^+, HCO_3^- and H^+) versus elements and chemical species whose sources could be different, including pollution (Cl^-, SO_4^-, NO_3^-). The second coordinate is an analysis inside the carbonatic and silicatic cycle, balancing cations and anions. The third compares the behaviour of the involved bivalent versus monovalent elements while the fourth and the fifth compare their relative behaviour. The sixth coordinate analyses the anions giving us information about the pH water conditions. Finally, the remaining coordinates investigate the behaviour of variables whose source may be related to pollution. Considering Cl^- in absence of atmospheric cyclic salts and evaporates about 30% of its amount is related to pollution, 54% in case of SO_4^{2-}, while for nitrate the most important anthropogenic sources are septic tanks, application of nitrogen-rich fertilizers to turf grass, and intensive agricultural processes (Berner and Berner 1996; Liu et al. 2011; Mencíó et al. 2016).

As we can see variance is higher for the first balance comparing natural and anthropic processes, and the last one, comparing SO_4^{2-} and NO_3^- whose ratio variability is a further witness of the presence of numerous sources/fluctuations. A first result here reveals that when elements are more related to natural weathering processes their balance variability appears to be reduced, probably indicating that the same processes have been working through time in a similar way. By taking into account the previous discussion about the dynamics of geochemical systems more information should be obtained by the analysis of the frequency distribution of the balances.

To achieve this aim in Fig. 16.3 an improved version of the *balance dendrogram* is reported where the original boxplots (Pawlowsky-Glahn and Egozcue 2011) are associated with the frequency distribution of the coordinates. Histograms have the same horizontal and vertical scale so they are comparable. Red line is related to the Gaussian distribution, black treated line to the Kernel density estimation.

Application of several normality tests indicates that under no circumstances the Normal distribution can be considered as model for the log-ratio coordinates; the consequence is that the log-normal model cannot be used to describe ratios between parts or group of parts. In most of the cases it appears to be due to some bimodality or to the presence of a heavy tail in the right-hand part of the distribution. The presence of power laws is associated with complex systems composed of processes that interact to self-organize their behavior across multiple temporal and/or spatial scales. Both fractals and multifractals are commonly associated with local self-similarity or scale-independence, generally leading to power-law relations

Table 16.1 Sequential binary partition process for the groundwater chemistry of the Arezzo basin (Tuscany, central Italy). Units of the chemical concentrations are mol/L while the variance expresses the contribution of each balance in explaining the total variability (% in parenthesis)

Balance	mol_Ca	mol_Mg	mol_Na	mol_K	mol_HCO$_3$	mol_Cl	mol_SO$_4$	mol_NO$_3$	mol_H$^+$	Variance
ilr$_1$	1	1	1	1	1	−1	−1	−1	1	1.23 (20.6)
ilr$_2$	1	1	1	1	−1	0	0	0	−1	0.29 (4.86)
ilr$_3$	1	1	−1	−1	0	0	0	0	0	0.54 (9)
ilr$_4$	−1	1	0	0	0	0	0	0	0	0.18 (3)
ilr$_5$	0	0	1	−1	0	0	0	0	0	0.94 (15.7)
ilr$_6$	0	0	0	0	1	0	0	0	−1	0.23 (3.85)
ilr$_7$	0	0	0	0	0	1	−1	−1	0	0.99 (16.6)
ilr$_8$	0	0	0	0	0	0	1	−1	0	1.57 (26.3)

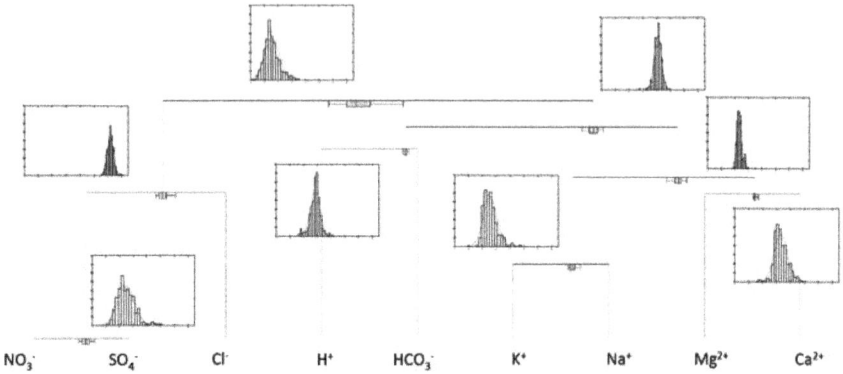

Fig. 16.3 Balance dendrogram (Thió-Henestrosa et al. 2008) with associated histograms. Red line corresponds to the Gaussian model, black treated line to the Kernel density estimation. The length of the vertical bar represents the proportion of the sample total variance

(Agterberg 2014). On the other hand the lognormal shape represents a special condition in which the interdependencies among processes are minimized or absent and repeated fragmentation (or dilution) dominates. As we can see in Fig. 16.3 the presence of heavy tails characterizes coordinates that mainly balance weathering of silicate and carbonates (K^+, Na^+, Mg^{2+}, Ca^{2+}, H^+, HCO_3^-) versus other environmental processes (NO_3^-, SO_4^-, Cl^-). Moreover, considering the internal partition of the previous balances, K^+/Na^+, Mg^{2+}/Ca^{2+} and, in particular, NO_3^-/SO_4^- ratios repeat this type of behavior.

The use of the complementary distribution function reveals the presence of power laws more clearly. In this plot, reported in Fig. 16.4, if X has a power law distribution the behavior of the *Prob[X ≥ x]* will be a straight line (Mitzenmacher 2004). As we can see, linear models can well describe several portions of curves for all the coordinates. This condition asks for multifractality perhaps associated to the space-time heterogeneity of the aquifer structure. Here a sudden change in the number of data with given concentration values is expected, particularly for pollution processes (Agterberg 2014). The fractal dimension of the phenomena, related to the slope of the straight lines, indicates how much more often there are low differences between the data rather then high differences.

On the whole the aquifer system appears to be governed by an interaction-dominant dynamics but it does not present a clear multimodality (or bimodality) that could be associated to different states. By considering Fig. 16.1 and the information deduced by the shape of the frequency distribution (Figs. 16.3 and 16.4) the aquifer could be associated with a sufficient resilience and recovery state (Scheffer et al. 2009, 2012). Of notice here is that the most important contribution to variability appears to be related to chemicals such as NO_3^- and SO_4^- suggesting the weight and the intermittency of the anthropic pressure. The multifractality revealed in Fig. 16.4 could indicate that in the dynamical system the energy

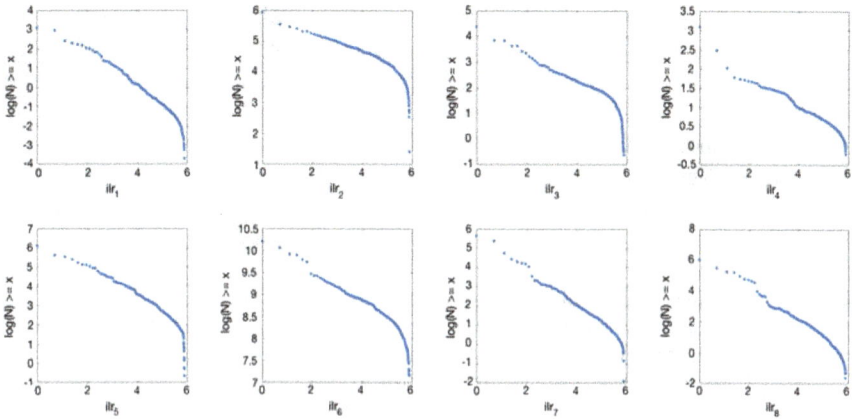

Fig. 16.4 Complementary distribution function to reveals the presence of power laws. If X has a power law distribution the behavior of the $Prob[X \geq x]$ will be a straight line (Mitzenmacher 2004)

dissipation cannot be neglected and that extended areas (intervals) of low fluctuations intermittent with small areas of extremely large fluctuations are to be expected. Moreover, the system as a whole is undergoing a non-linear dissipation with the energy interchange on different scales.

16.4 Conclusions

Starting from Garrels and Christ (1965) equilibrium in the water-rock system is usually analysed through the application of thermodynamic methods. In this context the statistical analysis of water concentrations, opportunely transformed into isometric logratio coordinates, could be an effective approach to understand where the randomness in nature comes from (Agterberg 2014) and if equilibrium conditions are really encountered.

The frequency distribution of the ratio of the compositional parts of Arezzo aquifer chemistry exhibits an overlapping between log-normal and power-law probability distributions when silicate and carbonate weathering (K^+, Na^+, Mg^{2+}, Ca^{2+}, H^+, HCO_3^-) is balanced versus other environmental processes (NO_3^-, SO_4^-, Cl^-). Similar results are obtained when the partition to generate new balances is applied to the previous group of parts (NO_3^- versus SO_4^-, K^+ versus Na^+ or Mg^{2+} versus Ca^{2+}). The result indicates a system subjected to nonlinear compositional changes due to presence of feedback effects attributable in a porous medium to change in porosity causing a remarkable change in permeability, in the pore-fluid flow and in the chemical-species concentration (Zhao 2014). Since thermodynamic equilibrium represents a homogeneous distribution of the parts, the obtained results indicate that the system is able to create and maintain a given amount of gradient, generating heterogeneity. However no clear multimodality is present and for the span of time here analysed different steady states (basins of attraction for concentration values) have not yet clearly emerged. Thus, from a compositional point of view, the system could be characterised by sufficient resilience and recovery rate from disturbances since the dissipative behaviour appears to be able to adsorb fluctuations. New progress would be made in this direction by exploiting the capacity of CoDA to capture the interdependence of concentration values, thus describing the water system and the surrounding as a whole, as in reality.

Acknowledgements This research was supported by the University of Florence (2016 funds) and by the GEOBASI project financed in 2015 by Tuscany Region through CNR-IGG. Frits Agterberg and B.S. Daya Sagar are warmly thanked for their support as well as IAMG for the 2003 Felix Chayes Prize and the constant sustain to my research activity. Lisa Merli helped me for the English language.

References

Aitchison J (1986) The statistical analysis of compositional data. In: Monographs on statistics and applied probability. Chapman and Hall, Ltd, London, UK, 416 pp (reprinted 2003 with additional material by The Balckburn Press)

Agterberg F (2014) Geomathematics: theoretical foundations, applications and future developments. Quantitative geology and geostatistics series, vol 18. Springer, 553 pp

Allegre CJ, Lewin E (1995) Scaling laws and geochemical distributions. Earth Planetary Sci Lett 132:1–13

Berner EK, Berner RA (1996) Global environment. In: Water, air, and geochemical cycles. Prentice Hall, Upper Saddle River, New Jersey, 07458, 376 pp

Billheimer D, Guttorp P, Fagan W (2001) Statistical interpretation of species composition. J Am Stat Assoc 96(456):1205–1214

Buccianti A, Magli R (2011) Metric concepts and implications in describing compositional changes for world river's chemistry. Comput Geosci 37(5):670–676

Buccianti A, Pawlowsky-Glahn B (2005) New perspectives on water chemistry and compositional data analysis. Math Geol 37(7):703–727

Buccianti A, Zuo R (2016) Weathering reactions and isometric log-ratio coordinates: do they speak each other? Appl Geochem 75:189–199

Buccianti A, Nisi B, Raco B (2016) Towards the concept of background/baseline compositions: a practicable path? In: Springer Proceedings in Mathematics and Statistics, vol 187, pp 31–43

Buccianti A, Lima A, Albanese S, DeVivo B (2018) Measuring the change under compositional data analysis (CoDA): insight on the dynamics of geochemical systems. J Geochem Explor 189:100–108

Cardenas MB (2008) Surface water-groundwater interface geomorphology leads to scaling of residence times. Geophys Res Lett 35(L08402):1–5

Cardenas MB, Jiang XW (2010) Groundwater flow, transport, and residence times through topography-driven basins with exponentially decreasing permeability and porosity. Water Resour Res 46(W11538):1–9

Chayes F (1960) On correlation between variables of constant sum. J Geophys Res 65(12): 4185–4193

Dai L, Korolev KS, Gore J (2015) Relation between stability and resilience determines the performance of early warning signals under different environmental drivers. PNAS 112(32): 10056–10061

Dakos V, Carpenter SR, van Nes EH, Scheffer M (2014) Resilience indicators: prospects and limitations for early warnings of regime shifts. Philos Trans B 370(20130263):1–10

Egozcue JJ, Pawlowsky-Glahn V, Mateu-Figueras V (2003) Isometric logratio transformations for compositional data analysis. Math Geol 35(3):279–300

Egozcue JJ, Pawlowsky-Glahn V (2005) Groups of parts and their balances in compositional data analysis. Math Geol 37(7):795–828

Engle MA, Rowan EL (2013) Interpretation of Na-Cl-Br systematics in sedimentary basin brines: comparison of concentration, element ratio, and isometric log-ratio approaches. Math Geosci 45(1):87–101

Engle MA, Blondes MS (2014) Linking compositional data analysis with thermodynamic geochemical modeling: oilfield brines from the Permian Basin, USA. J Geochem Explor 141:61–70

Engle MA, Rowan EL (2014) Geochemical evolution of produced waters from hydraulic fracturing of the Marcellus Shale, northern Appalachian Basin: a multivariate compositional data analysis approach

Fisher NI (1995) Statistical analysis of circular data. Cambridge University Press, UK, p 277

Garrels RM, Christ CL (1965) Solutions, minerals and equilibrium. Harper and Row, NY

Holden JG, Rajaraman S (2012) The self-organization of a spoken word. Front Psychol 3(209):1–24

Kleidon A (2010) Life, hierarchy, and the thermodynamic machinery of planet Earth. Phys Life Rev 7:424–460

Liu Z, Dreybrodt W, Liu H (2011) Atmosperic CO_2 sink: silicate weathering or carbonate weathering? Appl Geochem 26:S292–S294

Menció A, Mas-Pla J, Otero N, Regàs O, Boy-Roura M, Puig R, Bach J, Domènech C, Zamorano M, Brusi D, Folch A (2016) Nitrate pollution of graoudwater; all righ…, but nothing else? Sci Total Environ 539:241–251

Mitzenmacher M (2004) A brief history of generative models for power law and lognormaldistributions. Internet Math 1(2):226–251

Nicolis G, Prigogine I (1989) Exploring complexity: an introduction. Freeman and Company, New York, p 343

Owen DDR, Pawlwosky-Glahn V, Egozcue JJ, Buccianti A, Bradd JM (2016) Compositional data analysis as a robust tool to delineate hydrochemical facies within and between gas-bearing aquifers. Water Resour Res 52(8):5771–5793

Pawlowsky-Glahn V, Buccianti A (2011) Compositional data analysis: theory and applications. Wiley Ltd, 378 p

Pawlowsky-Glahn V, Egozcue JJ (2001) Geometric approach to statistical analysis on the simplex. Stoch Environ Res Risk Assess (SERRA) 15(5):384–398

Pawlowsky-Glahn V, Egozcue JJ (2011) Exploring compositional data with the CoDa dendrogram. Austrian J Stat 40(1/2):103–113

Pawlowsky-Glahn V, Egozcue JJ, Tolosana-Delgado R (2015) Modeling and analysis of compositional data, statistics in practice series. Wiley Ltd, 247 pp

Pearson K (1897) Mathematical contributions to the theory of evolution. On a form of spurious correlation which may arise when indices are used in the measurement of organs. Proc R Soc Lond, LX, 489–502

Seely AJ, Macklem P (2012) Fractal variability: an emergent property of complex dissipative systems. Chaos 22:0131081–013108-7

Scheffer M, Bascompte J, Brock WA, Brovkin V, Carpenter SR, Dakos V, Held H, van Nes EH, Rietkerk M, Sugihara G (2009) Early-warning signlas for critical transitions. Nature 461:53–59

Scheffer M, Carpenter SR, Lenton TM, Bascompte J, Brock W, Dakos V, van de Koppel J, van de Leemput IA, Levin SA, van Nes EH, Pascual M, Vandermeer J (2012) Anticipating critical transitions. Science 338:344–348

Shelton JL, Engle MA, Buccianti A, Blondes M (2018) The isometric log-ratio (ilr)-ion plot: a proposed alternative to the Piper diagram. J Geochem Explor 190:130–141

Shvartsev SL (2009) Self-organizing abiogenic dissipative structures in the geologic history of the earth. Earth Sci Front 16(6):257–275

Shvartsev SL (2013) Water-rock interaction: implications for the origin and program of global evolution. Int J Sci 2(10):26–30

Thió-Henestrosa S, Egozcue JJ, Pawlowsky-Glahn V, Kovacs LO, Kovacs G (2008) Balance dendogram, a new routine of codapack. Comput Geosci 34(12):1682–1696

van Rooij MMJW, Nash BA, Rajaraman S, Holde JG (2013) A fractal approach to dynamic inference and distribution analysis. Front Physiol 4:1–16

Zhao C (2014) Physical and chemical dissolution front instability in porous media. Lecture notes in earth system sciences. Springer International Publishing Switzerland, 354 pp

17

Soil Geochemical Analysis in USA

E. C. Grunsky, L. J. Drew and D. B. Smith

Abstract A multi-element soil geochemical survey was conducted over the conterminous United States from 2007–2010 in which 4,857 sites were sampled representing a density of 1 site per approximately 1,600 km^2. Following adjustments for censoring and dropping highly censored elements, a total of 41 elements were retained. A logcentred transform was applied to the data followed by the application of a principal component analysis. Using the 10 most dominant principal components for each layer (surface soil, A-horizon, C-horizon) the application of random forest classification analysis reveals continental-scale spatial features that reflect bedrock source variability. Classification accuracies range from near zero to greater than 74% for 17 surface lithologies that have been mapped across the conterminous United States. The differences of classification accuracy between the Surface Layer, A- and C-Horizons do not vary significantly. This approach confirms that the soil geochemistry across the conterminous United States retains the characteristics of the underlying geology regardless of the position in the soil profile.

E. C. Grunsky (✉)
Department of Earth and Environmental Sciences, University of Waterloo,
Waterloo, ON, Canada
e-mail: egrunsky@gmail.com

E. C. Grunsky
China University of Geosciences, Beijing, China

L. J. Drew
United States Geological Survey, Reston, VA, USA

D. B. Smith
United States Geological Survey, Denver, CO, USA

17.1   Introduction

A continental-scale soil geochemical survey was conducted over the conterminous United States from 2007 to 2010 by the U.S. Geological Survey (Smith et al. 2011, 2012, 2013, 2014). The survey collected samples at 4857 sites (Fig. 17.1), representing a density of 1 site per approximately 1600 km^2. The sampling protocol included, at each site, a sample from a depth of 0–5 cm (referred to as the surface soil for the remainder of this paper), a composite of the soil A horizon (the uppermost mineral soil), and a sample from the soil C horizon (generally the partially weathered parent material). If the top of the C horizon was at a depth greater than 1 m, a sample over a 20 cm interval was collected at a depth of approximately 1 m.

Studies on the geochemistry of two transects (east-west and north-south) across the United States and Canada, conducted as pilot studies in preparation for the continental-scale survey (Smith 2009; Smith et al. 2009) showed variability of soil geochemistry and mineralogy along both directions (Garrett 2009; Eberl and Smith 2009; Woodruff et al. 2009). As well, Drew et al. (2010) studied the two transects and demonstrated that the geochemical variability of soil is also closely associated with ecoregions (CEC 1997), which reflect continental scale features such as soil, landform, major vegetation types and climate. These studies indicate that the soil geochemistry is useful for mapping both geological and ecological domains.

Soil geochemistry, from a geological context, reflects a range of mineralogy, as a function of weathering of different parent materials, along with organic content due to biological activity. Ideally, soil geochemistry will represent underlying parent material and processes associated with the modification of those parent materials through comminution, weathering, ground water activity and biogenic processes. Grunsky et al. (2012, 2014) smf Mueller and Grunsky (2016) demonstrated that the

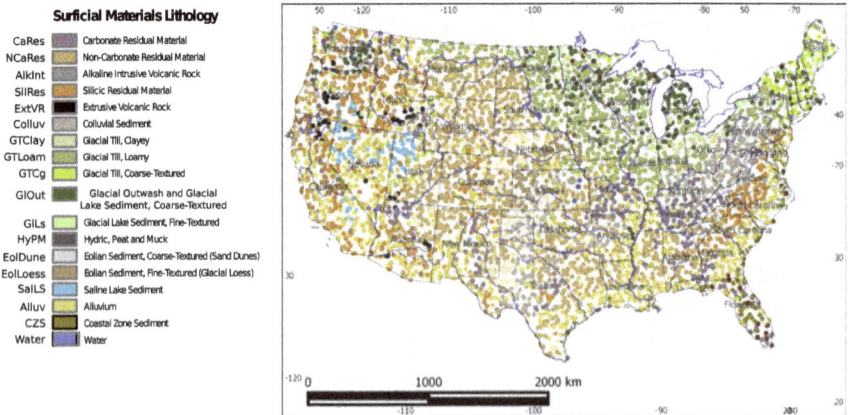

Surficial Materials Lithology

CaRes	Carbonate Residual Material
NCaRes	Non-Carbonate Residual Material
AlkInt	Alkaline Intrusive Volcanic Rock
SilRes	Silicic Residual Material
ExtVR	Extrusive Volcanic Rock
Colluv	Colluvial Sediment
GTClay	Glacial Till, Clayey
GTLoam	Glacial Till, Loamy
GTCg	Glacial Till, Coarse-Textured
GlOut	Glacial Outwash and Glacial Lake Sediment, Coarse-Textured
GlLs	Glacial Lake Sediment, Fine-Textured
HyPM	Hydric, Peat and Muck
EolDune	Eolian Sediment, Coarse-Textured (Sand Dunes)
EolLoess	Eolian Sediment, Fine-Textured (Glacial Loess)
SalLS	Saline Lake Sediment
Alluv	Alluvium
CZS	Coastal Zone Sediment
Water	Water

Fig. 17.1 Soil sample sites over the conterminous United States. Samples were taken at the (0–5) cm layer, the A- and C-horizons

geochemistry of lake sediment and glacial till in northern Canada can be used to predict the underlying lithologies. As part of the North American Soil Geochemistry Landscape Project (Smith et al. 2009), Grunsky et al. (2013) used soil geochemistry collected over the Maritime Provinces of Canada and the northeast United States to demonstrate that A-, B- and C-horizon soils geochemistry is useful for mapping the underlying lithologies. More recently, Grunsky et al. (2017) have shown that geochemistry of surficial soils can identify and classify underlying crustal blocks across the Australian continent, even after extended periods of weathering, transport and reworking.

The approach is based on the use of training sets of representative lithologies. Unfortunately, there are no continental-scale lithologic maps or representative training sets which can be used for predictive bedrock lithologic mapping in Canada or the United States. Sayre et al. (2009) classified the land surface of the conterminous United States according to surficial materials lithology, terrestrial ecosystems and isobioclimate. Isobioclimatic zones were subdivided into thermotypes, (temperature) and ombrotypes (moisture). It follows that soil geochemistry is a proxy for processes controlled by climatic factors. A key question that arises from this is can any of these processes be identified uniquely in the soil geochemistry and, if so, how can these processes be identified in terms of spatial continuity and distinctive chemistry? Drew et al. (2010) studied two transects across the US and demonstrated that the soil geochemistry is closely tied to zones that define the terrestrial ecosystems intersected by these transects. The objective of the current study is to address this question through the use of multivariate statistical analysis and Bayesian-based classification in conjunction with geostatistical methods that accurately describe processes in terms of distinctive geochemistry and spatial continuity.

17.2 Methods

17.2.1 Sampling and Analysis

The soil samples were analysed for geochemistry and mineralogy as described by Smith et al. (2011, 2012, 2013, 2014). The samples were air-dried and sieved to <2 mm after which the material was crushed in a ceramic mill prior to chemical analysis. Concentrations of Ag, Al, Ba, Be, Bi, Ca, Cd, Ce, Co, Cr, Cs, Cu, Fe, Ga, In, K, La, Li, Mg, Mn, Mo, Na, Nb, Ni, P, Pb, Rb, S, Sb, Sc, Sn, Sr, Te, Th, Ti, Tl, U, V, W, Y, Zn in all the soil samples (14,434) were determined using a near-total digestion using HCl-HNO_3-$HClO_4$-HF followed by inductively coupled plasma-mass spectrometry and inductively coupled plasma-atomic emission spectrometry. Mercury values were obtained using cold-vapor atomic absorption spectrometry following dissolution in a mixture of HCl and HNO_3 and Se was determined by hydride-generation atomic absorption spectrometry (HGAAS) following dissolution in a mixture of HNO_3, HF, and $HClO_4$. Arsenic was also determined by HGAAS following fusion in a mixture of sodium peroxide and sodium hydroxide at 750 °C. Total carbon was determined by combustion. Smith et al. (2013) provides details on the analytical methods and quality control protocols. Silicon was not determined.

All A-horizon and C-horizon samples (9575) were analysed by X-ray diffraction, and the percentages of major mineral phases were calculated using a Rietveld refinement method. Splits of the <2 mm fraction were used for analysis. Complete details of the technique and quality control protocols are provided in Smith et al. (2013).

17.2.2 Data Screening and the Compositional Nature of Geochemical Data

Geochemical analyses require screening and adjustment prior to any application of statistical methods and interpretation. A generalized sequence of data screening and adjustment strategies is documented in Grunsky (2010). The data were evaluated and analysed using the R programming and statistical environment (R Core Team 2013).

Major element concentrations, reported as percentages, were converted to ppm, by multiplying the values by a factor 10,000. Summary statistics for the data are given in Smith et al. (2013). The data were screened to determine the number of values that were reported at less than the lower limit of detection. Data that are reported at less than the lower limit of detection are termed as "censored". Censored data, when used in the application of statistical procedures, can influence estimates of mean and variance and therefore a replacement value that accurately reflects an estimate of the true mean is preferred. Furthermore, geochemical data are, by definition, compositions and as such the issue of closure becomes important (Aitchison 1986). Egozcue et al. (2003) describe various transformations that assist in evaluating data that are constrained by the effect of closure. For censored geochemical data, replacement values can be determined using the several methods based on maximum likelihood estimates of replacements values (Palarea-Albaladejo et al. 2014). Elements in which >80% of the values were censored were dropped from further evaluation, which included Ag, Cs and Te.

The data were also screened for sample sites where a large number of elements were reported at less than the lower limit of detection (<LLD). In the surface soil, 8 sites were found to have more than 25 elements reported at <LLD (3 from Florida). For the A horizon, 2 sites, all from Florida, were found to have more than 25 elements reported at <LLD. For the C horizon, 3 sample sites, in Florida, were found to more that have more than 25 elements reported at <LLD. These sites were dropped from further evaluation.

Summary statistics for the elements are provided by Smith et al. (2013, 2014). The remaining 43 elements: Al, As, Ba, Be, Bi, total C, Ca, Cd, Ce, Co, Cr, Cs, Cu, Fe, Ga, Hg, In, K, La, Li, Mg, Mn, Mo, Na, Nb, Ni, P, Pb, Rb, S, Sb, Sc, Se, Sn, Sr, Th, Ti, Tl, U, V, W, Y, Zn were then evaluated for the estimate of replacement values for those results that were reported at less than the lower limit of detection. The method of nearest neighbour replacement estimates (R package: zCompositions, function **lrEM**) was used on the censored data (Palarea-Albaladejo et al. 2014). The adjusted data were then used for subsequent multivariate statistical analysis.

17.2.3   *Integration of Land Surface Parameters with Soil Geochemistry*

Land surface maps of the conterminous United States (Sayre et al. 2009) were used to test the effectiveness of the soil geochemistry for revealing information on surficial materials lithology, terrestrial ecosystems and isobioclimate. Isobioclimatic zones were subdivided into thermotypes, (temperature) and ombrotypes (moisture). In this study, only the surface lithologies were studied in further detail. The results of the evaluation of the soil geochemistry in the context of terrestrial ecosystems, thermotypes and ombrotypes will be provided at a later time.

The maps were obtained as raster images with a pixel resolution of 1 km and a geodetic projection of decimal degrees using the North American Datum of 1983 (NAD83). These images were re-projected to the Lambert Conformal Conic projection using the following parameters (Spheroid—GRS 1980; Central Meridian: 96° West; Standard Parallels of 32° and 44°; Latitude of Origin: 38°; False Eastings and Northings of 0 m). This projection was used throughout the study.

The Quantum Geographic Information Systems (QGIS) (QGIS Development Team 2016) was used for the integration of various data sources and the geospatial rendering of the results. Within QGIS, two procedures were used from the Geospatial Data Abstraction Library (GDAL) procedure, "**warp (reprojection)**" and "**point sampling tool**". The map images were initially re-projected to the Lambert Conformal Conic (**lcc**) projection listed above using the "**warp**" procedure. The point dataset of the geochemical sampling sites were also reprojected from latitude/longitude coordinates to the **lcc** projection. The **lcc** image of the surface lithology was then sampled at the geochemical site coordinates using the "**point sampling tool**" and the surface lithology value was integrated into the geochemical database. This methodology was carried out for the other land surface maps (terrestrial ecosystems, surface lithologies, thermotypes and ombrotypes). The values of these features were integrated into the soil geochemistry dataset for further evaluation. It should be noted that the maps produced by Sayre et al. (2009) are generalizations and expressed at a resolution of 30 m (landforms, topographic moisture), 1 km (biogeographic regions) and 15 km for the surface lithology. It is possible that the class defined at any given point on the maps produced by Sayre does not correspond with the surface lithology, biogeographic, landform or topographic classes that were encountered during the soil survey sampling program.

For geospatial rendering purposes (interpolation), the Level 1 Ecology map of the conterminous United States was used to create a grid with a cell size of 40 km × 40 km.

Interpolation of principal component scores, posterior probabilities and measures of typicality were carried out using a geostatistical framework. The gstat package (Pebesma 2004) was used to generate and model semi-variograms with sufficient parameters to generate interpolated images through kriging. The cell size used for image interpolation was chosen as 40 km, the approximate spacing of the site sampling locations.

17.2.4   Process Discovery—Empirical Investigation of Soil Geochemistry

After screening the data for detection limit issues and missing values, the geochemical data were then subjected to an empirical investigation in which the assumptions about the data are minimal. To deal with the effect of closure, the data for 41 elements (Al As Ba Be Bi Ca Cd Ce Co Cr Cu Fe Ga Hg In K La Li Mg Mn Mo Na Nb Ni P Pb Rb S Sb Sc Se Sn Sr Th Ti Tl U V W Y Zn) were log-centred transformed after which a principal component analysis (PCA) was carried out using the methodology of Zhou et al. (1983) and Grunsky (2001). PCA was carried out on the entire set of multi-element data for the surface soil, the A and C horizons combined. PCA was also carried out on the multi-element data individually for the surface soil, A and C horizons. The rationale for this is based on enhancement of the multi-element signature for each layer rather than a principal component signature derived from the combined layers. The principal component biplots and corresponding maps of the component scores were subsequently generated for the surface soil, the A- and C-horizons independently. The biplots and interpolated maps provide insight into the orthogonal linear relationships that can reflect dominant geochemical processes that are influenced by mineral stoichiometry. The three soil layers were evaluated together in order to show any possible relationships between the two soil horizons (A and C) and the surface soil layer. To assist with insight into processes that influence the relationship of the elements and patterns of the scores of the observations, the loadings of the elements were coloured according to the classification of Goldschmidt (1937) into lithophile. siderophile or chalcophile affinity Elements associated with the atmophile affinity were not considered in this study.

17.2.5   Process Validation—Modelled Investigation of Soil Geochemistry

Using the classified information derived from the land surface maps of Sayre et al. (2009), the geochemical data were used to establish the ability to predict these classifications using a cross-validation approach in which the data are repeatedly sub-sampled as part of the classification process.

Previous studies (Grunsky et al. 2012, 2014) demonstrated that the use of multivariate statistical methods was able to classify bedrock lithologies based on lake sediment and glacial till geochemical data using discriminant analysis. The methodology employed the results of principal component analysis (described above), followed by an analysis of variance and the application of linear discriminant analysis (Venables and Ripley 2002) to determine which principal components were best at classifying and predicting the bedrock lithologies. This approach relies on having a sufficient number of degrees of freedom and homogeneity of covariance between the classes of the training sets. An alternative to linear discriminant analysis is quadratic discriminant analysis (Venables and Ripley 2002), which compensates for the classes where the condition of homogeneity of covariance cannot be met. The results of applying these methods includes measures of posterior probability in which each site is assigned a measure of probability of belonging to each of the classes and the class with the highest posterior probability is assigned to that site. Posterior probabilities are also compositions, as the sum of the probabilities for all of the classes for each site must sum to 1.0 and are, therefore, compositional in nature.

Both methods were tested for discriminating between the surface lithologies in this study. However, a comparison of results between linear discriminant and quadratic discriminant analysis showed little difference in the results and some classes had to be omitted because of an insufficient number of training sites.

To overcome some of the problems of applying classification methods in previous studies, we employed the statistical method, Random Forests (Breiman 2001) as employed by Harris and Grunsky (2015) and used as part of a remote predictive mapping strategy (Harris et al. 2008). The Random Forest method is based on the construction of classification trees (Venables and Ripley 2002, Chap. 9) in which nodes (splits in classes) are based on continuous variables from which a series of branches in the tree will correctly classify (categorical variables) all of the data. The Random Forest method "grows" many trees and each tree provides a classification. Each classification is termed a vote and a classification is assigned to the forest with the most votes. A useful description of the methodology is provided in Breiman and Cutler (2016). The function "**randomForest**", herein referred to as "RF", from the package **randomForest** (Breiman and Cutler 2016) was used for the analysis.

For each tree that is created, a training set of approximately one-third of the data is drawn, with replacement and are left out of the sample population. This is known as the out-of-bag (oob) data and is used to get a running unbiased estimate of the classification error, as trees are added to the forest. Variable importance is also determined from the out-of-bag data. For each tree, all of the data are applied to the tree and "proximities" are determined for each pair of cases. If two cases occur at the same node, then the proximity of that pair is increased by one. When all of the trees have been estimated, the proximities are normalized by dividing by the number of trees. These proximities can be used for replacing missing data, identifying outliers and creating lower dimensional views of the data. Each tree is constructed from bootstrapping the original sample population and about one third of the data are left out from each bootstrap sample and not used in tree construction but are then classified from the tree created from the other two thirds of the sample population. An unbiased estimate of the classification error is determined from each case that is oob and did not classify correctly. Variable importance is determined by comparing oob classification results and the non-oob classification results after random permutations of each of the variables. Another measure of variable importance is determined by the Gini measure that is determined by the number of splits that are made for a given variable over all of the trees in the forest. Variables do not need to be pre-selected using techniques such as analysis of variance as the RF procedure determines which variables are the best classifiers.

Maps of the normalized votes, which are equivalent to posterior probabilities, can be created using geostatistical methods such as kriging. However, since the posterior probabilities are compositions and sum to 1.0, these values must be logratio transformed, followed by subsequent co-kriging, and then back transformed for subsequent geographic rendering (Pawlowsky-Glahn and Egozcue 2015; Mueller and Grunsky 2016). Instead, maps of the posterior probabilities for each of the classes were created by posting the sample sites with points and colours. An alternative to this would be to consider the un-normalized (raw) votes as independent and carry out kriging on these estimations. The results of these interpolations are provided in the Supplementary Annex.

17.3 Results

17.3.1 Process Discovery—Principal Component Analysis

A logcentred transform was applied to the adjusted data after which a principal component analysis was carried out. An examination of an ordered plot of eigenvalues in the form of a screeplot (Jolliffe 2002) are shown in Fig. 17.2a–d for (a) all of the data, (b) Surface Soil, (c) A horizon only and (d) C horizon only. Figure 17.2a–d display two important inflection points; at PCs 3 and 9. The first three eigenvalues define the dominant structure in the data and the next 5 display lesser but significant structure also. This is also expressed numerically in Table 17.1 where the first 10 eigenvalues are listed along with the associated cumulative contribution to the structure in the data. As shown in the screeplots of Fig. 17.2, a comparison of the first four successive eigenvalues between the C-horizon,

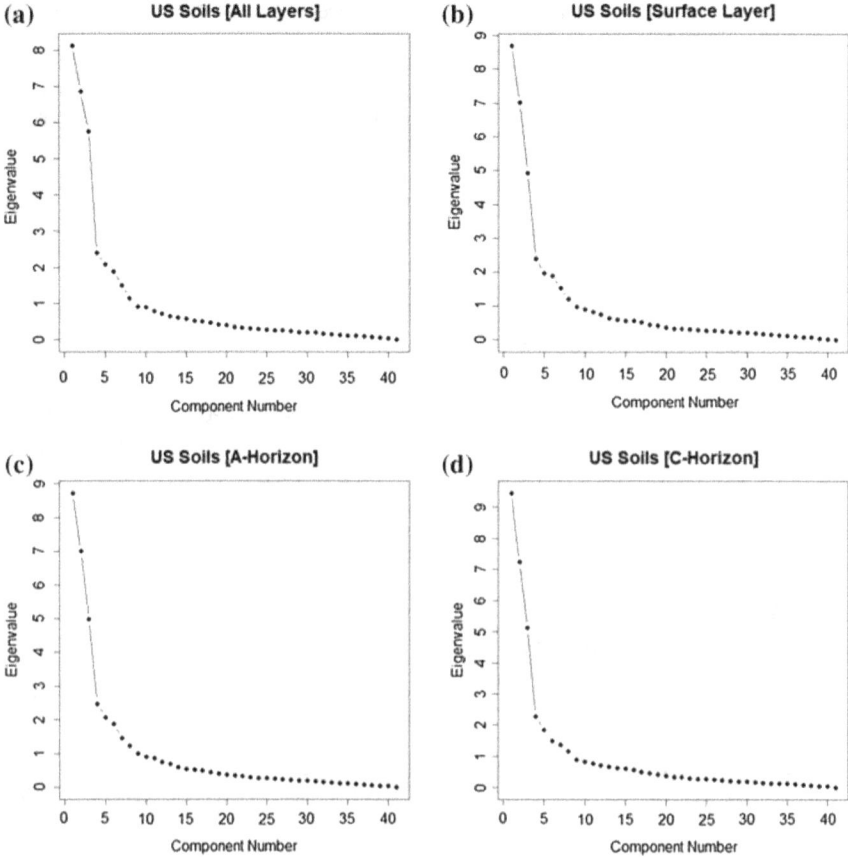

Fig. 17.2 a—Screeplot of eigenvalues of the soil geochemistry for the combined Surface Soil (0–5) cm layer, the A- and C- horizons, from the application of a principal component analysis to logcentred transformed data. b—Screeplot of eigenvalues of the soil geochemistry for the Surface Soil (0–5) cm layer from the application of a principal component analysis to logcentred transformed data for the top layer only. c—Screeplot of eigenvalues of the soil geochemistry for the A-horizon from the application of a principal component analysis to logcentred transformed data for the A-horizon only. d—Screeplot of eigenvalues of the soil geochemistry for the C-horizon from the application of a principal component analysis to logcentred transformed data for the C-horizon only

A-horizon and Surface Soil is slightly greater for the C-horizon. This implies that the linear combinations of the elements are stronger for the C-horizon than for the other two. Eigenvalues with values less than 1 and are interpreted to represent under-sampled processes or random effects (noise).

The largest eigenvalues signify that the linear combinations of the elements for these components are significant and defines "structure" in the data. This structure can be interpreted as the influence of stoichiometric control of mineralogy.

Table 17.1 Principal Component Analysis results for logcentred transformed soil geochemistry

RQPCA [clr] All layers

	PC1	PC2	PC3	PC4	PC5	PC6	PC7	PC8	PC9	PC10
λ	8.13	6.87	5.76	2.39	2.08	1.88	1.50	1.15	0.92	0.89
λ%	19.83	16.76	14.05	5.83	5.07	4.59	3.66	2.80	2.24	2.17
Σλ%	19.83	36.59	50.63	56.46	61.54	66.12	69.78	72.59	74.83	77.00

RQPCA [clr] surface soil

	PC1	PC2	PC3	PC4	PC5	PC6	PC7	PC8	PC9	PC10
λ	8.70	7.01	4.93	2.41	1.96	1.89	1.53	1.21	0.98	0.90
λ%	21.19	17.08	12.01	5.87	4.77	4.60	3.73	2.95	2.39	2.19
Σλ%	21.19	38.27	50.28	56.15	60.93	65.53	69.26	72.20	74.59	76.78

RQPCA [clr] A horizon

	PC1	PC2	PC3	PC4	PC5	PC6	PC7	PC8	PC9	PC10
λ	8.73	7.00	4.97	2.47	2.07	1.88	1.45	1.22	1.00	0.90
λ%	21.29	17.07	12.12	6.02	5.05	4.59	3.54	2.98	2.44	2.20
Σλ%	21.29	38.37	50.49	56.51	61.56	66.15	69.68	72.66	75.10	77.29

RQPCA [clr] C horizon

	PC1	PC2	PC3	PC4	PC5	PC6	PC7	PC8	PC9	PC10
λ	9.45	7.22	5.12	2.29	1.84	1.50	1.36	1.17	0.89	0.82
λ%	23.02	17.59	12.47	5.58	4.48	3.65	3.31	2.85	2.17	2.00
Σλ%	23.02	40.61	53.08	58.66	63.14	66.80	70.11	72.96	75.13	77.13

17.3.2   PCA of the Combined Surface Soil, A-Horizon, C-Horizon

Figures 17.3a, and 17.4a shows biplots (PC1-PC2 and PC2-PC3) for the principal component scores and loadings for the combined data from the surface soil, A- and C-horizons Table 17.1 shows that the first three principal components for the combined data (All Layers) account for 50.6% of the overall variation in the data.

Figure 17.3a shows the mass of data points defined by two vertices: (1) Cr-V-Ni-Co-Fe-Sc-Mn-P-Zn; (2) Hg-In-Ti-Se-Mo-As-Sb-Sn-Bi (chalcophile) and a trend of element associations: Mg-Ca-Na-Sr-Ba-K-Be-Rb-Tl that are inversely associated with the vertex defined by (2) above. The chalcophile elements are grouped along the +PC1 axis. Siderophile elements are associated with the +PC2 axis and the lithophile elements are distributed around the ±PC1/−PC2 axes and the −PC1/+PC2 axes.

Figure 17.4a shows the three sets of data (Surface Layer, A- and C-horizon) combined onto a biplot of PC2–PC3. The PC scores along the PC2 axis define a contrast between mafic (+ scores) and felsic (−scores) source material. Siderophile (Fe, Co, Ni), lithophile (Cr, V, Sc, Ti) and chalcophile elements (Cu, In) are associated along the +PC2 axis and lithophile elements (Rb, K, Tl, Ba, Th, La, Be, Ce) are concentrated along the −PC2 axis.

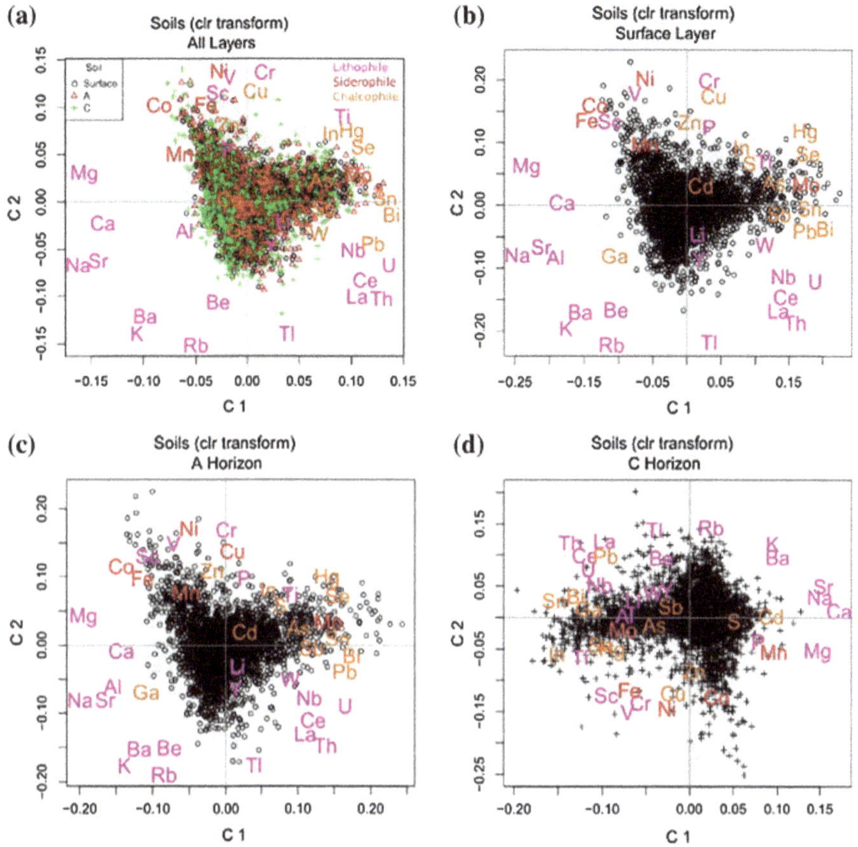

Fig. 17.3 a—Biplot of principal components 1 and 2 for the soil geochemistry for the combined Surface Layer, A, and C horizon soil geochemical data based on a log centred transform. The colours and symbols represent the surface soil and the soil A and C horizons. **b**—Biplot of principal components 1 and 2 for the Surface Soil geochemistry data based on a log centred transform. **c**—Biplot of principal components 1 and 2 for the A-horizon soil geochemistry data based on a log centred transform. **d**—Biplot of principal components 1 and 2 for the C-horizon soil geochemistry data based on a log centred transform

An association of chalcophile elements (Cd, S, Sb, As, Hg, Pb) occurs along the +PC3 axis with a corresponding concentration of sample sites associated with the surface layer and A-horizon, most likely representing complexing with organic rich soils. PC scores for the C-horizon are concentrated along the ±PC2 axis, which may represent a range of source material from mineral soils that are low in organic material (−PC3) to soils that are rich in organic material or derived from shales/weathered materials (+PC3).

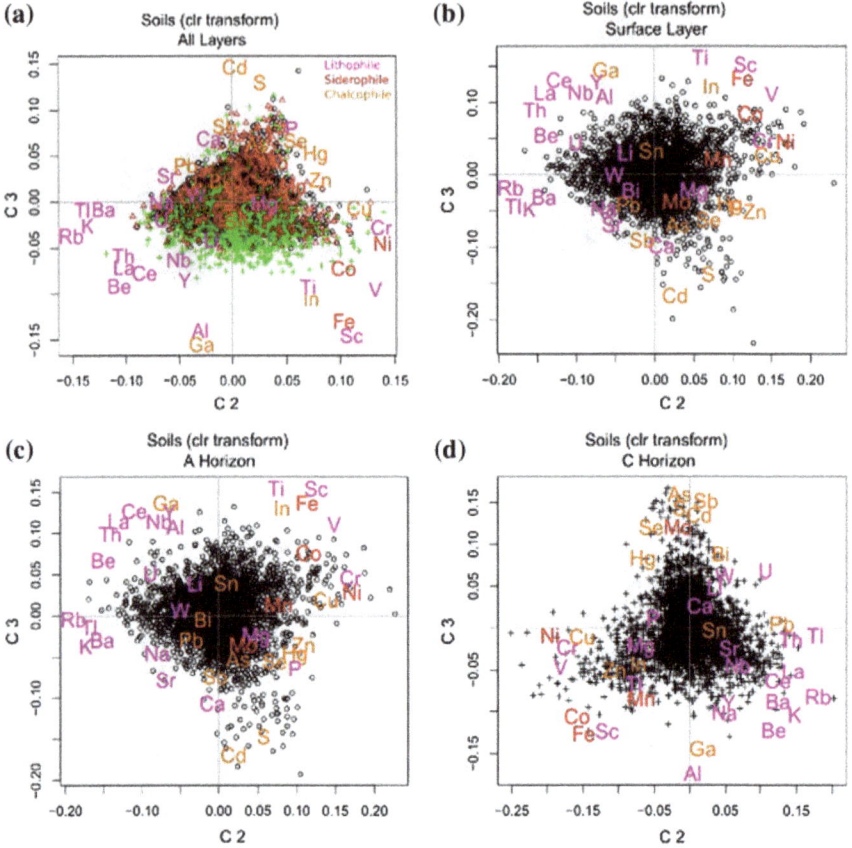

Fig. 17.4 **a**—Biplot of principal components 2 and 3 for the soil geochemistry for the combined Surface Soil, A, and C horizon soil geochemical data based on a log centred transform. The colours and symbols represent the surface soil and the soil A and C horizons as shown in Fig. 17.3a. **b**—Biplot of principal components 2 and 3 for the top layer soil geochemistry data based on a log centred transform. **c**—Biplot of principal components 2 and 3 for the A-horizon soil geochemistry data based on a log centred transform. **d**—Biplot of principal components 2 and 3 for the C-horizon soil geochemistry data based on a log centred transform

17.3.3   PCA of the Surface Soil, A-Horizon, C-Horizon

The biplots of Fig. 17.3a–c for all of the data, the surface soil data and the A-horizon data, show similar patterns in terms of the relationships of the elements with each other and the shape of the data cloud for the projection of the principal component scores onto the PC1 and PC2 axes. The biplots exhibit a range of lithophile loadings that define materials derived from mafic, feldspathic, carbonate and REE-enriched sources within the quadrants described previously. Similarly, the chalcophile element association is concentrated along the +PC1 axis for both

Fig. 17.3b, c, likely representing weathered and organic-rich material, which adsorb chalcophile elements.

The biplot of Fig. 17.3d (C-horizon) displays a different pattern in comparison with Fig. 17.3a–c. The +PC1 axis shows an association of lithophile elements (Ca-Mg-Na-Sr-P) and chalcophile elements (S-Cd), possibly representing a mix of feldspathic and/or carbonate source material. Along the PC1 axis and on the +PC2 domain, there is a contrast between (Ca-Na-Mg-S-Ba-K) and (Th-Ce-U-La-Nb-Al-Li) that may reflect a feldspathic/carbonate source environment from an environment with relative enrichment in heavy minerals.

Figure 17.4a shows a pattern and association of elements that displays a contrast of the C-horizon data with the surface soil and A-horizon data. Figure 17.4a shows a siderophile and mafic lithophile pattern of Cr-Ni-Cu-V-Co-Fe-Sc along the +PC2 axis. Along the −PC2 axis of Fig. 17.4a there is a lithophile association of Rb-K-Ti_Ba-Ce-La-Tl. The +PC3 axis in Fig. 17.4a shows a chalcophile/lithophile association of Cd-S-Sb-Ca-P-Se-Hg-As-Mo-Pb-Sr-Zn. This region of the plot is dominated by surface soil and A-horizon data although some C-horizon data are also present. A similar pattern is observed in Figs. 17.4b, c although the groups of the elements are at opposite ends of PC3 (a sign switch). In Fig. 17.4b, c, transitional between the siderophile/lithophile elements (Fe-Sc-Co-Cr-Ni) and the lithophile elements (Rb-Tl-K-Ba) is the grouping of Al-Ga-Nb-Y-Ce-La-Th-U that represents feldspars, clays and heavy minerals. As in Figs. 17.3d and 17.4d, representing the C-horizon data, shows the chalcophile enrichment trend along the +PC3 axis and a siderophile/lithophile trend along the PC2 axis. Transitional between the trend along the PC2 axis is an association of Al-Ga, likely representing feldspars and clays.

17.3.4 Mapping the Components

The first three principal components for the surface soil, the A- and the C-horizons were interpolated using the geostatistical package, gstat (Pebesma 2004). Experimental semi-variograms were generated followed by variogram model fitting with subsequent kriging. The images for the three principal components are shown in Figs. 17.5a–c, 17.6a–c and 17.7a–c.

Principal Component 1

Geospatially these patterns are observed in Figs. 17.5, 17.6 and 17.7. Figure 17.5a–c show interpolated images based on kriging of the first principal component for the surface soil, A- and C-horizons respectively. The patterns observed in Fig. 17.5a and b are consistent with the patterns observed in Fig. 17.3b and c. The +PC1 axis in Fig. 17.3b and c show relative enrichment of the previously identified chalcophile elements and relative enrichment of the mafic lithophile and siderophile elements along the −PC1 axis. In Fig. 17.5a and b, the positive scores of PC1 appear to correspond with the region in the southeast US and the negative scores of PC1

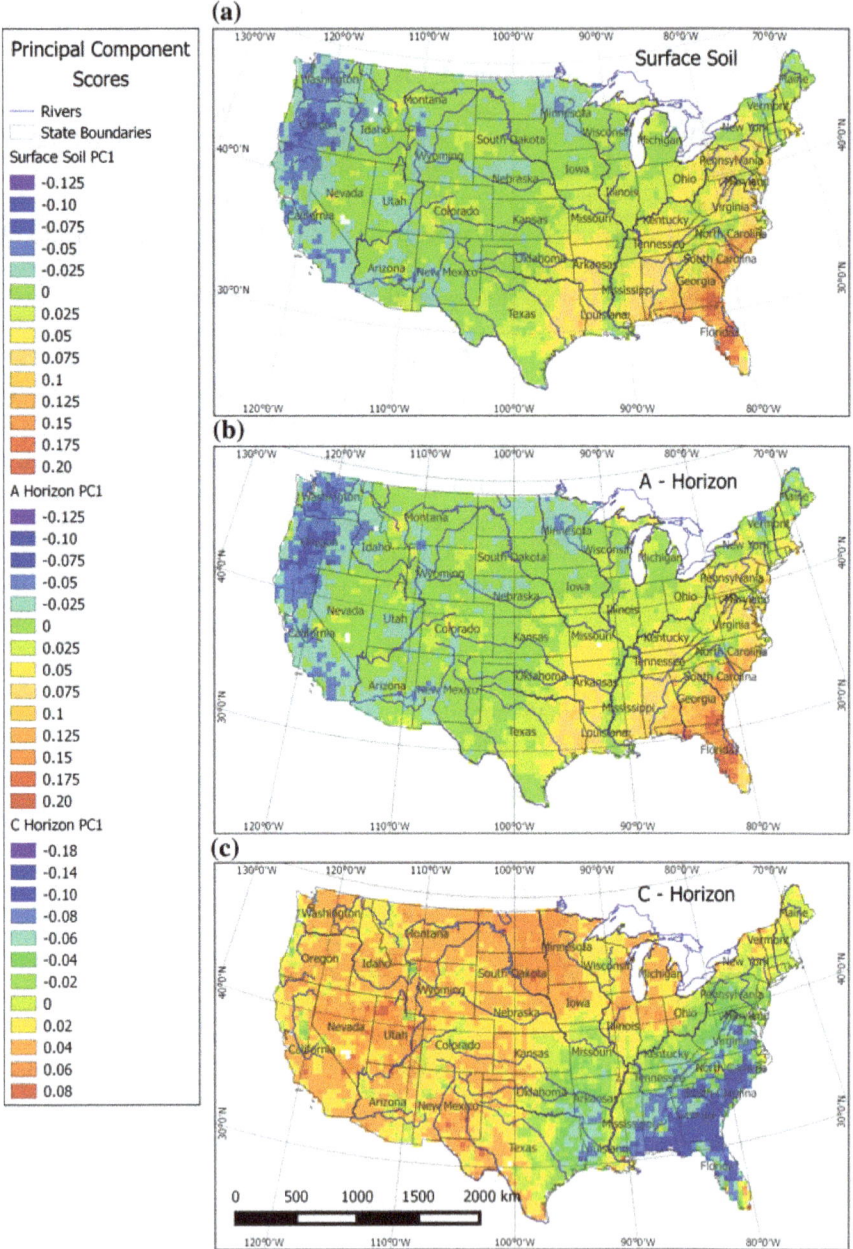

Fig. 17.5 a–c Map of kriged principal component 1 for the Surface Soil, A- and C-horizon data. Figures 17.4b–d provide the context for relative element enrichment/depletion associated with each of the layers

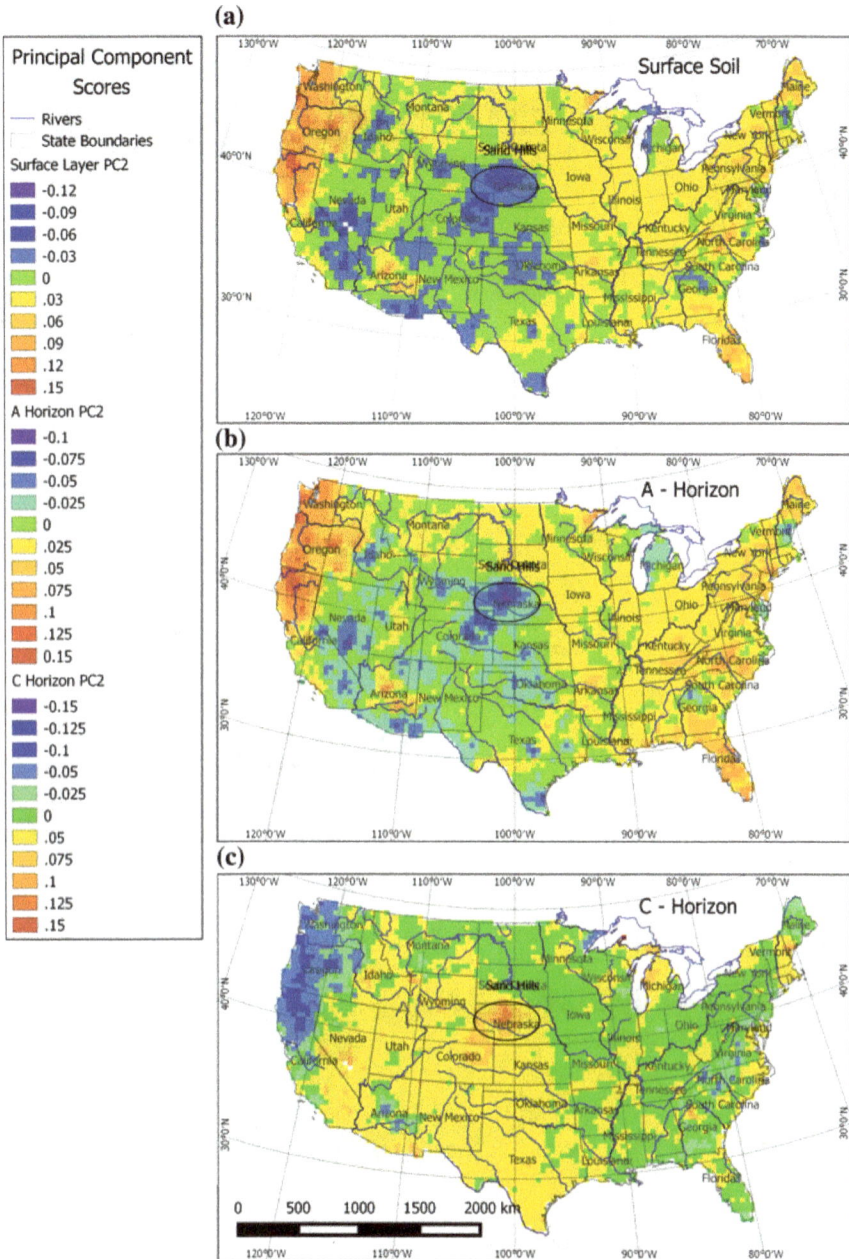

Fig. 17.6 a–c Map of kriged principal component 2 for the Surface Soil, A- and C-horizon data. Figures 17.4b–d provide the context for relative element enrichment/depletion associated with each of the layers

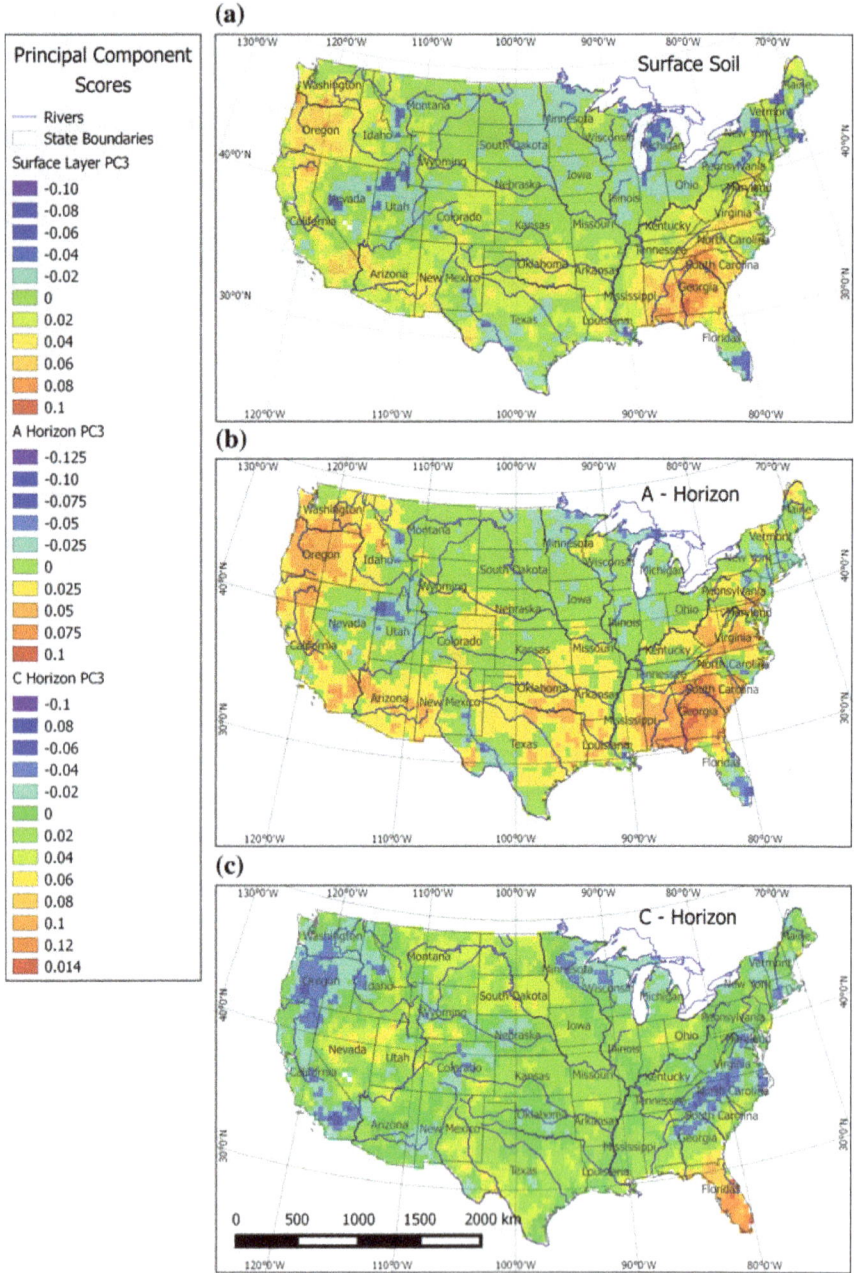

Fig. 17.7 a–c Map of kriged principal component 3 for the Surface Soil, A- and C-horizon data. Figures 17.4b–d provide the context for relative element enrichment/depletion associated with each of the layers

appear to occur in the northwest US and west of Lake Superior. All three figures show a pattern that coincides with the banks of the Mississippi River. Negative PC1 scores for the surface layer and A-horizon correspond to relative enrichment in Na-Sr-Al-Ca-Mg-K-Ba element associated with feldspars and/or carbonate source material.

The image of PC1 for the C-horizon data (Fig. 17.5c) shows a strong negative region in the southeast US that corresponds to the chalcophile group of elements along the negative portion of PC1 in the biplot of Fig. 17.3d. The positive portion of PC1 in Fig. 17.3d corresponds to the dominantly lithophile and siderophile groups of elements and is displayed as a large region throughout the US, with the exception of the southeast US. The same "corridor" pattern along the Mississippi River is observed in Fig. 17.5c, for the C-horizon results and represent the same relative concentration of lithophile elements observed in the surface layer and A-horizon.

Figure 17.5c shows the kriged image for the first principal component derived from the C-horizon data. In this case, the negative scores are restricted to the eastern US and reflect the chalcophile and rare earth elements indicative of detrital heavy minerals corresponding to the region of quartz enrichment accompanied with weathered and detrital materials within the erosional and weathering domain of the eastern US. Positive PC1 scores reflect a lithophile association of Ca-Na-Sr-Cd-Mg-Ba-K-Mn (Fig. 17.3d) and suggest an environment that is likely dominated by Ca-Na-K-Ba-Sr feldspars and Mg-Ca bearing ferromagnesian minerals.

An important consideration in the interpretation of the biplots is the significance of the associations of the elements. An initial interpretation of the biplots of Fig. 17.3a–d was that the associations of the chalcophile groups indicated relative enrichment of these elements (Hg-Se-As-Sb-Sn-Bi-Pb-S-In) that represent weathered materials along with the accumulation of detrital minerals within the erosional and weathering domain of the southeastern US. In fact, these elements do not reflect relative enrichment but rather relative depletion with respect to the other groups of elements, notably the siderophile and lithophile elements. Geospatially, the chalcophile association of these elements corresponds to the region of a high quartz content in the soil (Smith et al. 2014) and has been termed the "quartz dilution effect". This effect in the soil geochemistry and the subsequent multi-element associations would likely be significantly different had Si been included in the analysis. A test was carried out in which the Si content of the data was simulated as the difference from the potential total (1,000,000 ppm) from the summed content of the compositions. This simulated Si value was then included in the composition and a PCA was carried out. The first component identified the relative Si enrichment as occurring in the southeast US. The simulated value of Si was not included in this study because other elements should also be considered in a total composition, including oxygen and nitrogen.

Principal Component 2
As shown in Fig. 17.3b, c, the multi-element signature of tpc2 is nearly the same for the surface soil and A-horizon. The patterns in both figures show two trends, one

with relative enrichment in Cr-Ni-Co-Cu-V-Fe-Sc (siderophile/lithophile + Cu-Zn) and the other with relative enrichment in Hg-Se-As-Sn-Sb-Pb-Bi-In-S. (chalcophile) These two multi-element associations reflect the chemistry of mafic minerals and elements that are associated with weathering and organic complexing. This is reflected in the maps of Fig. 17.6a, b in which high PC2 values are noted in the eastern and south eastern US and the western US. The negative PC scores for the surface soil and A-horizon show relative enrichment in Rb-K-Tl-Ba-Be-Na-Sr-Al-Ga and, as shown in Fig. 17.6a, b are geospatially concentrated in the central US corresponding to the location of the Sand Hills of Nebraska, (~105° W/ 42° N), which is comprised of sand-sized particles of quartz and feldspar (Smith et al. 2014). There are also areas of negative PC2 scores, most likely representing feldspars associated with granitoid rocks in southern Nevada, California, Arizona, Texas, New Hampshire and Maine (Smith et al. 2014).

The map of PC2 (Fig. 17.6c) for the C-horizon data shows positive scores associated with the mafic volcanic rocks of the northwest US and corresponds to the relative enrichment of siderophile (Fe-Ni-Co), lithophile (Cr-V-Sc), chalcophile (Cu-Zn) elements as shown in Figs. 17.3d and 17.4d. The negative scores for PC2 show a similar pattern to those of the surface soil and A-horizon; relative enrichment in alkali lithophile elements (Rb-K-Ba-Be-Na-Sr) with Al-Ga representing feldspars and REE lithophile elements (U-Th-La-Ce-Ng-Tl) that represents heavy minerals and quartz (as explained previously). The geochemical expression of these minerals in PC2, which are resistant to weathering, are reflected in both horizons and the surface soil.

Principal Component 3

The positive scores for the PC3 show relative enrichment of siderophile, mafic lithophile, and light REE elements for both the surface soil and A-horizon; whereas this pattern is represented by negative scores for the C-horizon. As shown in Fig. 17.4b–d, for all three layers, there is a continual transition from relative enrichment in alkali lithophile and REE elements, including Al and Ga, representing feldspars and minerals associated with felsic domains to relative enrichment in Cr-Ni-V-Cu-C-Fe-Sc-Ti-In-Zn that represents minerals associated with mafic domains. Figures 17.7a–c show the kriged images for the third principal component. The negative scores show relative enrichment of Cd-S-Ca-Sr-Sb-P-As, which may reflect the processes of organic complexing and sulphates. Negative scores noted in Utah, Nevada, west Texas, the Mississippi delta and south Florida may have a greater component of S. Negative scores that occur in Minnesota, Michigan, Indiana and the coast of New England may reflect the presence of shales, clays and organic accumulations. The negative PC3 scores of Fig. 17.4b exhibit a bimodal pattern of relative enrichment of Fe-Sc-In-Ti and Ga-Al-Y-Nb-Ce-La. The Fe-rich pattern is associated with the mafic volcanic rocks in the northwest and southwest US and the Ga-rich pattern occurs in the eastern US and reflects the presence of feldspars in the weathering of granitoid rocks in the southern Appalachians.

As seen in Fig. 17.4c, and nearly identical to that the of surface soil, the positive scores of PC3 exhibit a bimodal pattern for the A-horizon and indicate relative enrichment of Ti-Sc-Fe-In-V and Ga-Al-Th-La-Nb-Ce. These two groups reflect both a mafic and feldspathic/heavy mineral rich environment. Figure 17.7b shows the mafic association (Ti-Sc-Fe-In-V) in the northwest US. The positive scores in the eastern, southern, and in particular, the southeast US reflect elements associated with feldspars and heavy minerals, which reflects the concentration of minerals through the weathering process, which may be due more to gravitational effects than chemical breakdown. As in Fig. 17.7a, the negative scores of PC3 in the A-horizon demonstrate the same patterns and processes.

The C-horizon map shows two distinct geospatial patterns. The positive scores of Fig. 17.4d show relative enrichment in the chalcophile group, Sb-As-S-Mo-Se-B-Cd-Hg-U-Li-W and occur primarily in the southeast US. This pattern likely reflects both the quartz dilution effect and the presence of chalcophile elements relative to other areas throughout the US. The negative scores, which show relative enrichment of the lithophile elements Al-Ga-Na-Y-K-Be-Ba-Mn-Ti-Fe-Sc-Co, reflect a combination of mafic minerals and feldspars. These patterns are observed in the western US, Minnesota-Wisconsin, central Appalachia and the northeast US. Patterns associated with the elements that reflect mafic domains are the northwest US and Wisconsin-Minnesota. Patterns that reflect the feldspathic domains are Nebraska-Colorado, central Appalachia and the northeast US.

Evaluation of the soil geochemistry for the surface soil, the soil A horizon and the soil C horizon using a principal component approach reveals that there are continental-scale geochemical patterns that appear to be associated with the composition of the underlying soil parent material, climate, and weathering. At the scale of evaluation, details on specific lithologies are difficult to resolve, but the patterns are consistent with those mineralogical patterns delineated by Smith et al. (2014).

Process Validation Predictive Mapping of Surface Lithologies

The lithology of surficial materials by Sayre et al. (2009) is represented by 18 classes plus unknowns and listed in Table 17.2. A total of 17 classes were selected for further study. The classes "unknown" and "water" were not used as they were not considered suitable for classification.

Figure 17.8 shows a map of the sampling sites with the surface materials lithology from Sayre et al. (2009). The patterns of surface materials on the map show some similarities with the patterns observed from the first three principal components for the surface soil, A- and C-horizons. Figure 17.9 shows a biplot of the first two principal components that are coded according to the surface lithologies. The pattern of the mafic lithophile elements (Cr-Ni-Cu-V-Co-Fe-Sc) in Fig. 17.9a, b are dominated by silica-rich residual soils (SilRes), whereas the chalcophile enrichment pattern (Hg-Se-Mo-Sn-Bi-Pb-Sb-As-Ti-S-In) appears to be associated mostly with alluvium (Alluv) and coastal zone sediments (CZS). The lithophile element grouping in the negative portion of the PC2 shows a mix of several lithologies. The results of the PCA suggest that the linear combinations of elements from the PCA are related to the patterns observed in Surface Materials Lithologies of Fig. 17.8.

Table 17.2 List of surface lithologies across the conterminous United States

Mnemonic	Class description	Surface layer	A-horizon	C-horizon	Total
AlkInt	Alkaline intrusive/volcanic rocks	6	7	6	19
Alluv	Alluvium and fine-textured coastal Zone sediment	994	989	984	2967
CaRes	Carbonate residual material	265	263	260	788
Colluv	Colluvial sediment	379	379	366	1124
CZS	Coastal zone sediment, coarse-textured	44	45	43	132
EolDune	Eolian sediment, coarse-textured (Sand Dunes)	152	151	151	454
EolLoess	Eolian sediment, fine-textured (Glacial Loess)	156	155	155	466
ExtVR	Extrusive volcanic Rock	50	51	51	152
GlLs	Glacial lake sediment, fine-textured	89	85	86	260
GlOut	Glacial outwash and Glacial lake sediment, coarse-textured	221	220	221	662
GTCg	Glacial till, coarse-textured	114	111	111	336
GTClay	Glacial till, Clayey	61	61	61	183
GTLoam	Glacial till, Loamy	529	528	526	1583
HyPM	Hydrick peat muck	25	25	26	76
NCaRes	Non-carbonate residual material	1188	1174	1170	3532
SalLS	Saline lake sediment	78	82	79	239
SilRes	Silicic residual material	457	456	452	1365
Water[a]		22	21	21	64
Unknown[a]		6	6	6	18
Total		4836	4809	4775	14420

[a]Not Used

From the application of the random forest classification, the Gini Index (significance of the variables) for the surface soil, A- and C-horizons are listed in Table 17.3 and shown graphically in Fig. 17.10. The significance uses the Gini Index, which is a measure of purity based on the success of a variable in distinguishing between classes. Table 17.3 shows that generally, PC's 4, 5, 1, 2, 3 and 6 are the best variables for classification of the surface lithologies for the surface soil, A- and C-horizons. Maps of the normalized votes in point form and interpolated (kriged) maps of the raw votes are shown in the Supplementary Annex (Supplementary Figs. 1–15).

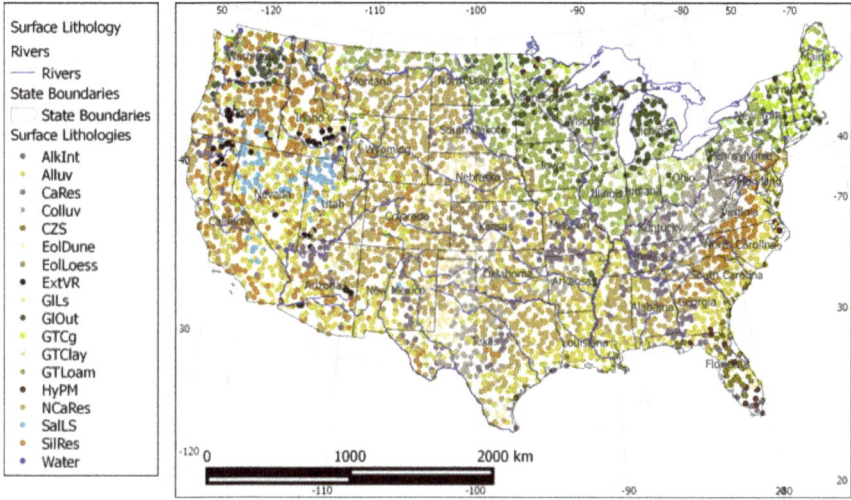

Fig. 17.8 Map of soil sample sites coded by the Surface Lithology classification. This map represents the actual classification based on the maps of Sayre et al. (2009). Colours used in this figure are the same colours used in Sayre's maps. See text for details on how the sites were selected

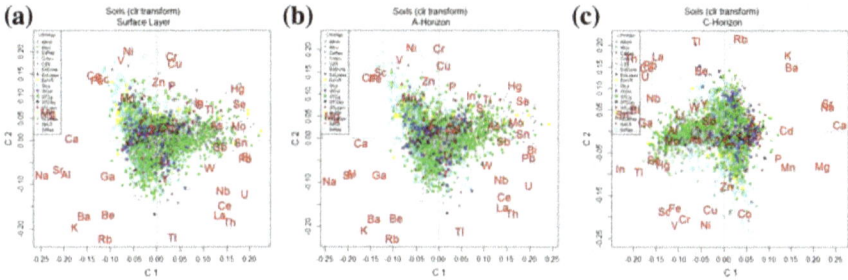

Fig. 17.9 **a–c** Principal component biplot of the surface layer (**a**), A-horizon (**b**) and C-horizon (**c**) scores that are coded and coloured according to the surface lithologies

Table 17.4 shows the accuracy of prediction for each of the surface lithologies based on the Random Forest out-of-bag classification methodology for each of the surface soil, A- and C-Horizons. The table has been ordered from the highest to the lowest prediction accuracies based on the surface soil. It is worth noting that the depth of soil has only a minor influence in the prediction accuracies, suggesting that the geochemical signature of the underlying material persists throughout the soil column. Non-carbonate residual soils (NCaRes) (~74%), loam associated with glacial till (GTLoam) (66–72%), siliceous residual soils (SilRes) (48–56%), alluvium (Alluv) (~50%) and coastal zone sediments (CZS) (45–48%) have the highest prediction accuracies, whereas the lowest accuracies are associated with hydric peat and muck (HyPM) (0%), alkalic intrusions (AlkInt) (0%), glacial lake sediments (GlLs)

Table 17.3 List of variable importance for the surface layer, A- and C-horizons as determined from Random Forest classification of the principal component results applied to the clr-transformed data. Colours reflect the most significant PCs (red) to least significant PCs (blue)

Surface Layer PC	Importance		A Horizon PC	Importance		C Horizon PC	Importance
4	198.35		4	185.34		2	165.83
5	180.88		5	172.04		4	156.36
1	163.70		1	170.09		1	154.76
2	155.81		3	150.05		3	152.11
3	152.73		2	148.14		6	131.94
9	129.17		9	127.18		16	128.21
12	109.74		6	126.50		5	115.55
6	108.61		20	119.91		11	113.04
32	106.28		29	110.74		8	109.54
23	102.15		13	102.50		7	109.05
30	100.84		12	100.25		10	106.46
8	98.87		11	98.07		14	101.91
20	98.77		8	96.86		13	96.87
40	98.19		23	94.89		12	96.57
10	96.90		10	94.48		28	95.32
21	95.83		7	93.99		9	94.97
11	93.68		16	92.35		18	93.88
15	91.76		18	92.21		17	92.57
24	91.73		19	91.46		31	92.53
13	91.49		21	89.23		34	92.34

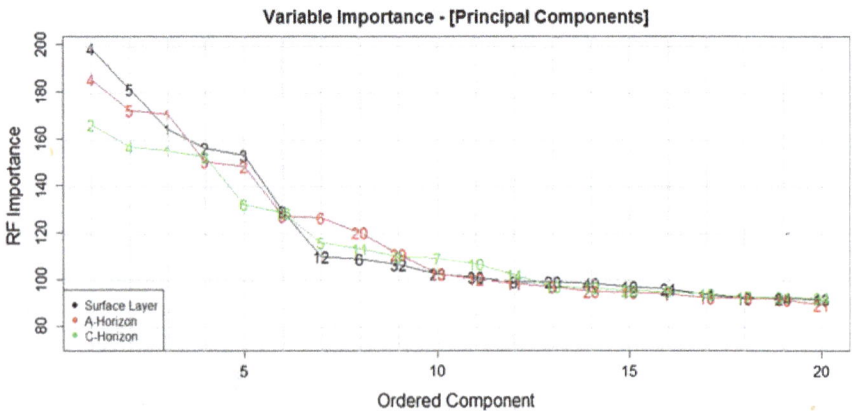

Fig. 17.10 Plot of the significance of the principal components used in the random forest classification based on the Gini Index for the Surface Layer, A- and C-horizons. See the text for a detailed explanation

Table 17.4 Measures of ordered predictive accuracy for the surface lithologies for the surface layer, the A- and C-horizons based on a Random Forest classification of the principal component results applied to the clr-transformed data

	Surface Layer	A-Horizon	C-Horizon
NCaRes	74.82	73.66	74.51
GTLoam	71.61	68.90	65.74
SilRes	52.02	47.97	55.92
Alluv	50.38	50.63	49.77
CZS	44.90	48.34	48.26
Colluv	37.14	38.20	32.73
GlOut	28.41	32.63	29.32
GTClay	27.54	37.32	42.23
GTCg	22.65	21.47	20.57
EolDune	22.25	22.40	16.47
EolLoess	21.05	26.97	30.19
CaRes	19.19	15.16	9.58
SalLS	15.22	15.69	1.25
ExtVR	1.96	5.78	0.00
GILs	1.11	0.00	0.00
AlkInt	0.00	0.00	0.00
HyPM	0.00	0.00	0.00
Overall Accuarcy	49.92	49.37	48.61

(0–1%) and extrusive volcanics (ExtVR) (0–6%). The prediction accuracy is sensitive to the initial representation of each class in the dataset. This sensitivity is partly due to the masking and swamping effect that a large population of sites for one type of surface lithology over another (i.e. Alluvium vs. Hydric Peat and Muck).

Supplementary Tables 2, 3 and 4 provide a complete summary of the prediction accuracies for the surface soil, A- and C-horizons, respectively. The diagonal of each upper table (Tables 2a, 3a, 4a) indicates how many sample sites were classified correctly. Each row of the off-diagonal elements indicates the misclassification of the sites for each of the classes. The lower tables in Tables 2b, 3b, 4b show the classification accuracies as expressed in percentages. The overall classification accuracy is shown at the bottom of each table. Scanning the columns of Tables 2a, 3a, and 4a reveals that many classes are confused with alluvium (Alluv), siliceous residual material (SilRes), loam derived from glacial till (GTLoam) and non-carbonate residual material (NCaRes). Alluvium and non-carbonate residual material appear to overlap with almost all of the classes. The overall prediction accuracies for the surface soil, A- and C-Horizons are 50%, 49% and 49%, respectively.

The R package "**randomForests**" produces raw and normalized votes for each of the classes. Votes are a record of the number of times a site is correctly classified. As described above, normalized votes are the equivalent of a posterior probability and are therefore compositions. Classes such as AlkInt, HyPM and other classes that have low abundance in the data create problems in the creation of co-regionalization that is required for co-kriging. Examples of the spatial distribution of the normalized and raw votes are shown below. The Supplementary Annex provides predictive maps for all of the surface lithologies, based on the normalized votes, for the surface soil, A- and C-horizons. Predictive maps for AlkInt and HyPM are not shown because the normalized votes for these two surface lithologies were very low and do not show any geospatial patterns. The prediction accuracies for the three media from Table 17.4 are: 49.9%, 49.4% and 48.6% respectively. Supplementary Tables 2, 3 and 4 provide details on the overlap of predictions for each surface lithology. In most cases, overlap is associated with non-carbonate residual soils, glacial till derived loam and alluvium. These three classes have the broadest range of compositional variation and occupy a significant amount of area across the conterminous US.

Figure 17.11 shows a map of normalized votes of Non-carbonate residual soils (NCaRes) derived from the random forest classification. Normalized votes >0.3 occur throughout the Midwest states from the Canadian border in the north to the Gulf of Mexico in the south. From Table 17.4, the overall classification accuracy is approximately 75% for the surface soil and the two soil horizons. Supplementary Tables 2, 3 and 4 show that compositional overlap occurs primarily with alluvium, which is also shown in the maps of Fig. 17.11 where a large number of sample sites show low normalized votes ($\sim$0.2–0.3). Supplementary Fig. 13a, b show the normalized and raw vote maps of the NCaRes prediction.

Figure 17.12 shows a map of normalized votes for loam derived from Glacial Till (GTLoam). The overall classification accuracy ranges from 65.7 to 71.6% over the three soil layers. Supplementary Tables 2, 3 and 4 show the overlap of the GTLoam composition is associated with non-carbonate residual material (NCaRes) and alluvium (Alluv) for the surface soil, A- and C-horizons (Supplementary Tables 2, 3, 4). The pattern of elevated normalized votes coincides with the region described by Sayre et al. (2009) that is located in the north central US and south of the Great Lakes. The pattern of elevated GTLoam follows the course of the Mississippi River, which highlights the erosional path of this material. Supplementary Figs. 12a, b show the normalized and raw vote maps of the GTLoam prediction.

Normalized votes for the prediction of alluvium (Alluv) are shown in Fig. 17.13 (Supplementary Fig. 1). The overall prediction accuracy is $\sim$50% (Table 17.4) and compositional overlap is observed with the surface lithology non-carbonate residual soil (NCaRes) (Supplementary Tables 2, 3, 4). High predictions of alluvium are located in Nevada, western Texas and the southeast US states. The dispersed prediction of 0.2–0.3 represents the regions of compositional overlap with NCaRes, which can be seen on the map of Fig. 8. Supplementary Figs. 1a, b show the normalized and raw vote maps of the Alluv prediction and supplementary Figs. 13a, b show the normalized and raw votes of the NCaRes prediction.

Figure 17.14 shows prediction based on the normalized votes for the Eolian Dunes (EolDune) of Nebraska, southward into Texas. The patterns are the same for the surface soil, A- and C-horizon maps. The highest values of normalized votes

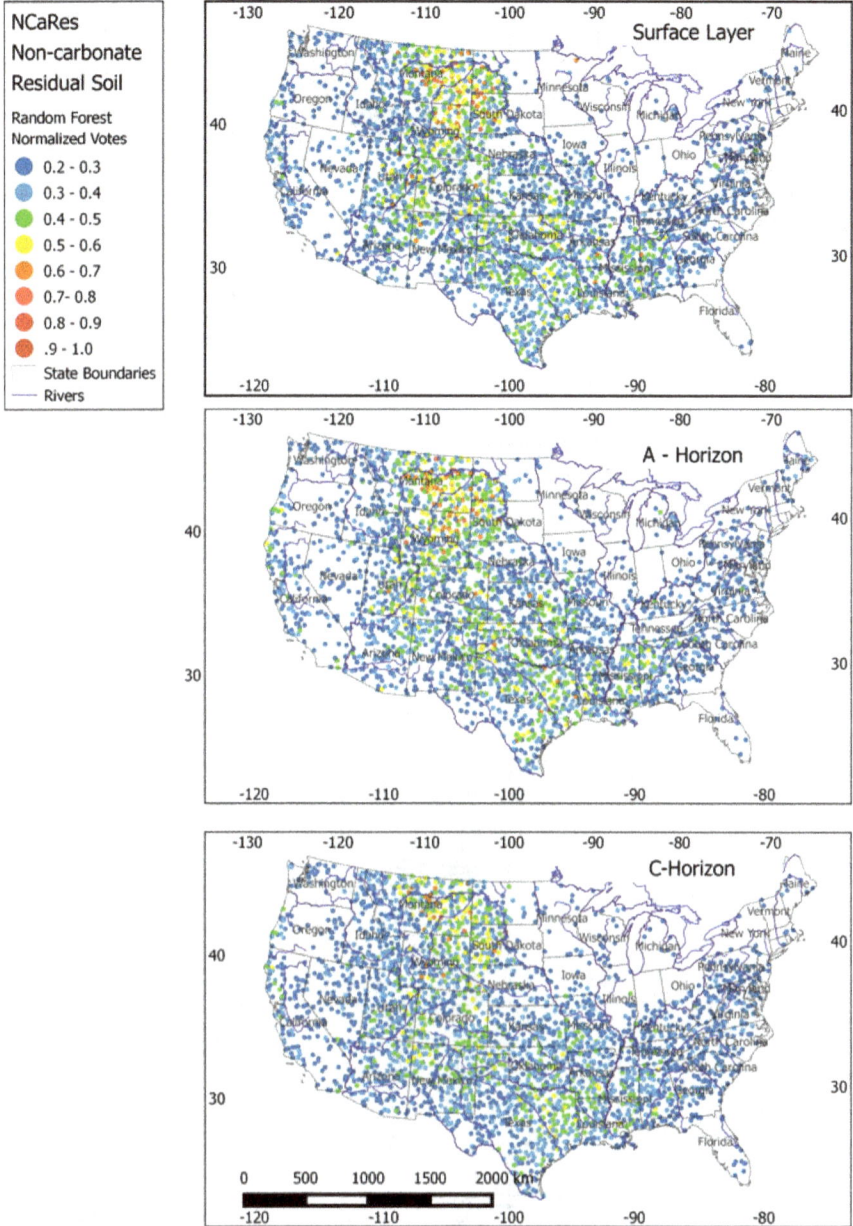

Fig. 17.11 Map of normalized votes for the surface lithology class, non-calcium residual soil (NCaRes). Sites with a normalized vote of less than 0.2 are omitted

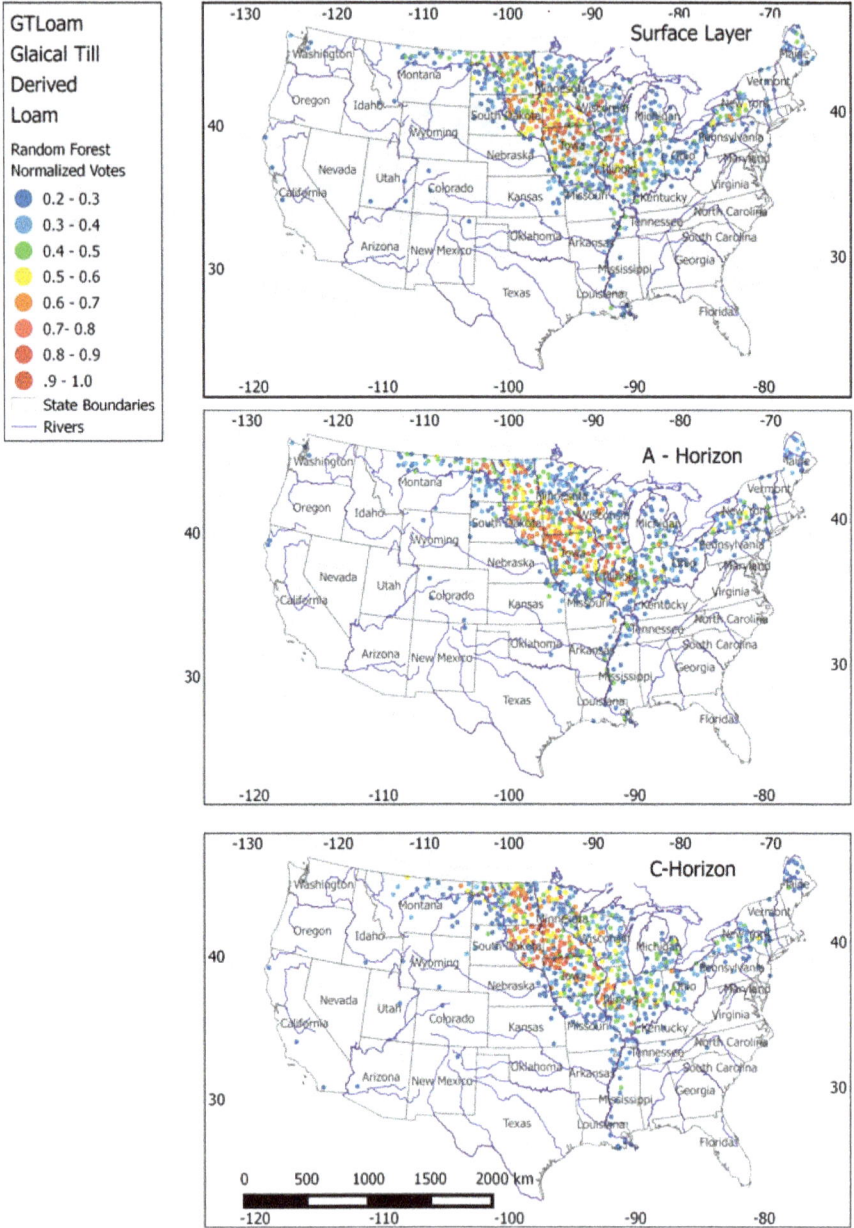

Fig. 17.12 Map of normalized votes for the surface lithology class, loam derived from glacial till (GTLoam). Sites with a normalized vote of less than 0.2 are omitted

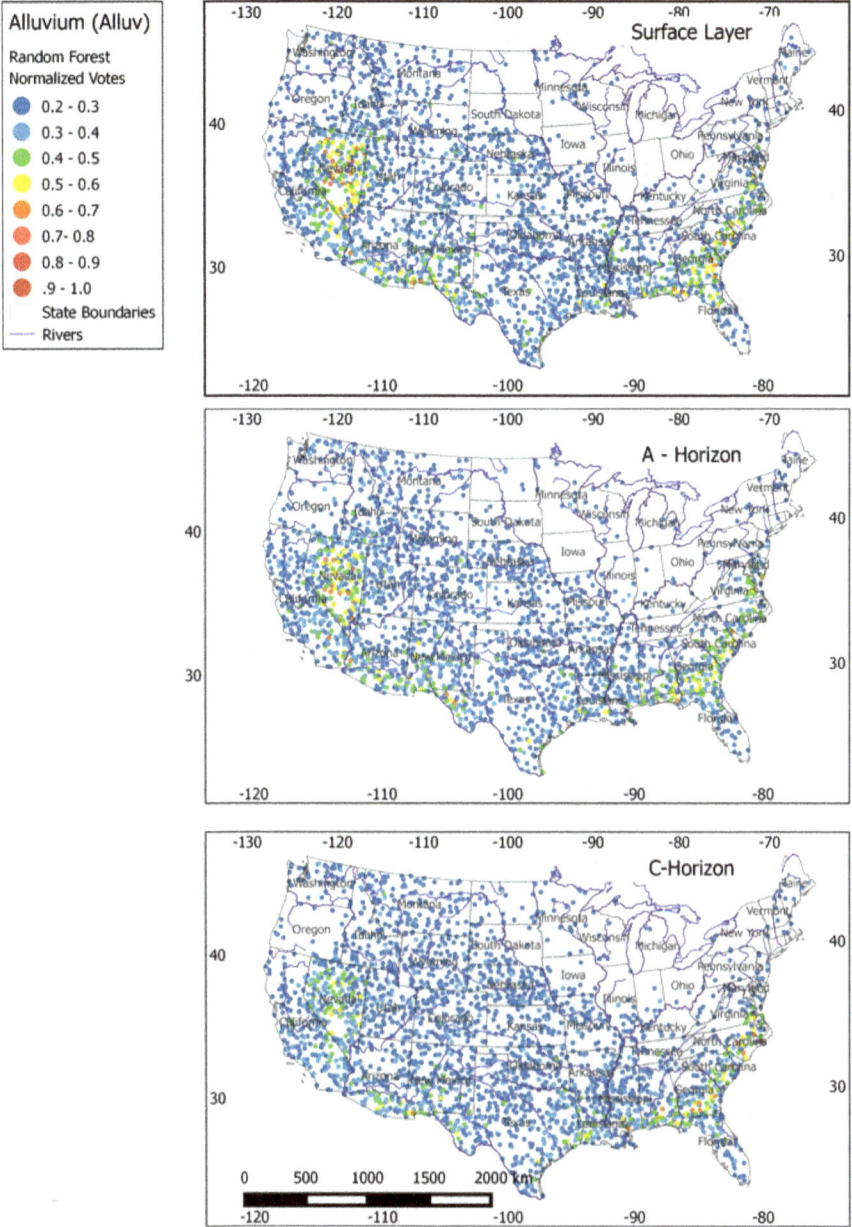

Fig. 17.13 Map of normalized votes for the surface lithology class, alluvium (Alluv). Sites with a normalized vote of less than 0.2 are omitted

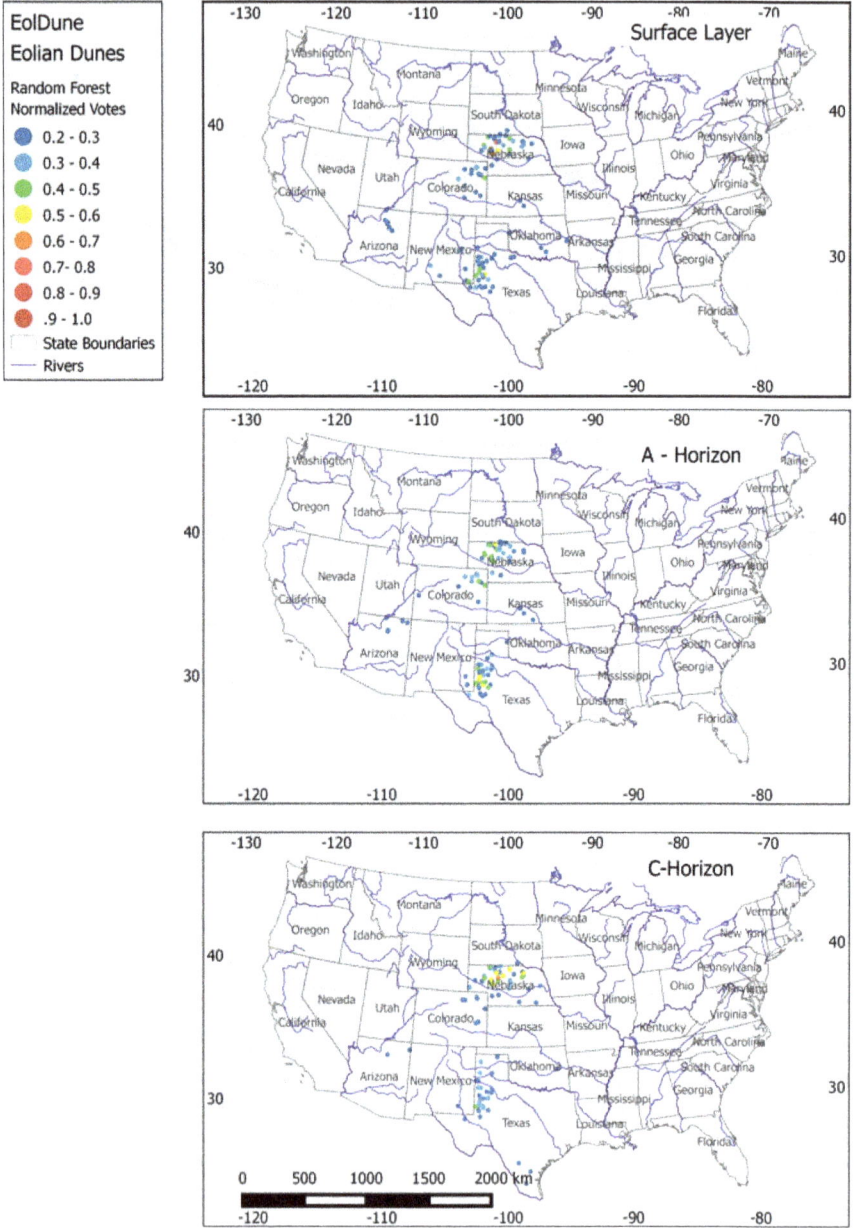

Fig. 17.14 Map of normalized votes for the surface lithology class, eolian dunes (EolDune). Sites with a normalized vote of less than 0.2 are omitted

occur in Nebraska and west-central Texas. The map of Sayre et al. (2009) shows EolDune in northern Texas and the Oklahoma Panhandle, although these two regions are not predicted in the surface soil, A- or C-Horizon results. Table 17.4 shows predictive accuracies of 22.3, 22.4 and 16.5% for the surface soil, A- and C-horizons, respectively. Supplementary Tables 2, 3 and 4 show that compositional overlap occurs with alluvium (Alluv) and non-carbonate residual soil (NCaRes). Supplementary Figs. 5a, b show the normalized and raw vote maps of the EolDune prediction.

The effects of erosion and subsequent re-deposition along the banks of the Mississippi River is observed for several of the surficial lithologies. NCaRes, CaRes and Colluv exhibit an erosional pattern along the Mississippi River, while EolLoess, GlLS, GlOut and GTLoam exhibit depositional patterns. This suggests that the recent deposition of the sediments along the banks of the Mississippi River has modified the composition of the upper layers of the soil. These classes (Eol-Loess, GlLS, GlOut, GTLoam—Supplementary Figs. 6a, b, 8a, b, 9a, b, 12a, b) show a distinct compositional presence down the length of Mississippi River starting from the northern Midwest states and reflecting continued transport of these materials at a continental scale.

A brief description of the maps for the surface soil, A and C-horizon data that are displayed in the Supplementary Annex are discussed in the section, Supplementary Material.

17.4 Discussion

Examination of the principal component biplots (Figs. 17.3 and 17.4) show that the multi-element patterns are very similar for the surface soil and A-horizon data. The C-horizon biplots show similar multi-element groupings, but the shape of the point patterns (Figs. 17.3d and 17.4d) are different from those of the surface soil and A-horizon (Figs. 17.3b, c and 17.4 b, c). As described previously, the element groupings for the three sampling layers are:

(1) Group 1: Tl-Rb-Be-Ba-K-Ga-Al-Sr-Na-Ca-Mg [felsic and mafic lithophile elements (silicates)]
(2) Group 2: Ni-Cr-V-Fe-Sc-Co-Cu-Zn-Mn [Ferromagnesian silicates and clays]
(3) Group3: Hg-Se-Mo-Sn-Bi-Pb-Sb-As-Ti-S-In. [Shales and organic material with adsorbed elements]

These associations are slight variants on Goldschmidt's classification of elements; lithophile (Group 1), siderophile (Group 2) and chalcophile (Group 3).

The principal component biplots, along with the maps of the dominant principal components (Figs. 17.5, 17.6 and 17.7), indicate that there is strong stoichiometric and geospatial control on the patterns that are observed. These patterns, both in the biplots and the kriged map images, provide the justification to use the soil geo-chemical data to predictively map (validate) the surface lithology classification of Sayre et al. (2009). It should be noted that Sayre's map of surface lithologies does not distinguish lithologies with different mineralogies, and, hence there is consid-erable overlap between some of the classes defined by Sayre.

The results of the random forest classification show that for most of the surface lithology classes, the accuracy of prediction and spatial coherence of the predicted sites is variable, as shown in Table 17.4 and Figs. 17.11, 17.12, 17.13 and 17.14 and the Supplementary Tables and Figures. The surface lithologies with the lowest predictions are: Hydric Peat and Muck (HyPM), Alkalic Intrusives (AlkInt), Glacial Lake Sediments (GlLS), Extrusive Volcanic Rocks (ExtVR) and Saline Lake Sediments (SalLS). Two factors influence the classification accuracy. The first is the areal extent that a given class occupies. The compositional range of a class of small spatial extent may be swamped or masked by the compositional range of a class that is geographically adjacent to it and has a much larger areal extent. Surface lithologies such as AlkInt, HyPM ExtVr, SalLS and GlLS have limited geospatial extent and the compositions of these lithologies are similar to several other lithologies, including Alluv GTLoam and NCaRes. The second factor that influ-ences the prediction accuracy is the common compositions of several of the surface lithology classes namely, alluvium (Alluv), non-carbonate residual soil (NCaRes), and silica-rich residual soil (SilRes). These surface lithologies are comprised of similar mineralogies and are, therefore, compositionally similar and result in compositional overlap in the statistically based prediction process.

Silicate mineralogy, including quartz, is under-represented in the data used for this study. As discussed previously, the quartz dilution effect has an influence on how the various relationships of the elements are observed, particularly in the methods that are part of the "Process Discovery" component of this study. The absence of silicon in the geochemical analysis in terms of the classifications may have some effect on the ability to distinguish between the different surface lithologies, but the exact effect is unknown at this time and further studies where Si is included and subsequently excluded in process discovery studies are warranted.

The validation of surface lithologies using soil geochemistry highlights some of the limitations on predicting distinct surface lithologies that have similar geo-chemical compositions but represent different processes. Despite this confusion of compositions between surface lithology classes, the predictive maps render a close representation of the maps of Sayre et al. (2009).

17.5 Concluding Remarks

The multi-element soil geochemistry over the conterminous United States contains a rich set of information that reflects the original source material and subsequent modification through weathering, mass transport, climate and biological activities.

As a result, continental-scale geochemistry may represent many processes. In this study, we have focused on the evaluation and interpretation of the multi-element soil geochemistry from the surface soil, A- and C-horizons in the context of predicting the surface lithologies.

Process discovery makes use of multivariate methods such as principal component analysis, which creates orthogonal linear combinations of the elements that often reflect processes controlled by mineral stoichiometry that comprise the parent material. This parent material may be bedrock (igneous, metamorphic, sedimentary), glacial deposits, loess or fluvial deposits. Ideally, soil geochemistry can be used to predict the composition of the underlying soil parent material. As demonstrated in this study, multivariate methods such as principal component analysis cannot decouple all of these processes. Processes such as igneous and metamorphic mineral reactions share similar mineral stoichiometry, making them indistinguishable from a geochemical perspective. Many distinct sedimentary assemblages are comprised of similar lithologies with similar mineralogy, and are thus difficult to distinguish solely on a geochemical basis.

With the exception of the surface lithology map of Sayre et al. (2009), a continental-scale map of lithology does not exist, which creates difficulty in an attempt to predictively map at large scales. However, the availability of the maps by Sayre et al. (2009) that include terrestrial ecosystems, thermoclimate, soil moisture and surface lithologies provides an opportunity to test the capacity of soil geochemistry to uniquely define these features. Although not presented here, the soil geochemistry has the ability to uniquely define terrestrial ecosystems and regional climate indicators. We intend to publish the results of using soil geochemistry to uniquely identify the terrestrial ecosystems, thermoclimatic zones and soil moisture (ombrotype) as defined by Sayre et al. (2009).

With few exceptions, there are only minor differences between the geochemical compositions of the surface soil and the A-horizon. The geochemistry of the C-horizon displays a distinct geochemical difference between the surface soil and A-horizon as it has not undergone the degree of weathering as the near-surface soils and contains less organic material.

The overall predictive accuracies for the predicting the surface lithologies for the surface soil, A- and C-horizons are 49.9%, 49.4% and 48.6%, respectively. As described above, the reasons for these low accuracies are due to the overlap of many of the lithologies with Alluvium, Non-carbonate residual soils, Siliceous soils, Eolian Dunes, Eolian Loess and materials deposited from glaciation. However, the spatial continuity of the posterior probabilities confirm the distinctiveness of these lithologies and demonstrate the effectiveness of soil geochemistry in recognizing the differences between the classes.

The geochemistry of soils represents modification of the initial parent material through weathering in response to varying precipitation and temperature, groundwater effects, meteoric water effects, biologic activity and geologic complexity.

Thus, geochemistry is a rich source of information that can be used in many ways to describe, monitor and predict processes derived from natural and anthropogenic events (Grunsky et al. 2013).

The results from the statistical evaluation of the geochemical data in the context of predicting surface lithologies across the conterminous US indicates that soil geochemistry reflects a number of physical processes. Further studies of the soil geochemistry across the US will evaluate the ability to predict terrestrial ecosystems and indicators of climate.

Acknowledgements The authors thank Karl Ellefson of the United States Geological Survey for his thoughtful and helpful review of the manuscript.

References

Aitchison J (1986) The statistical analysis of compositional data. Chapman and Hall, New York, p 416

Breiman L (2001) Randomforests. Machine. Learning 45:5–32

Breiman L, Cutler A (2016) Random forests. https://www.stat.berkeley.edu/~breiman/RandomForests/cc_home.htm#intro

Commission for Environmental Cooperation (CEC) (1997) Ecological regions of North America—toward a common perspective. Montreal: commission for Environmental Cooperation, p 71

Drew LD, Grunsky EC, Sutphin DM, Woodruff LG (2010) Multivariate analysis of the geochemistry and mineralogy of soils along two continental-scale transects in North America. Sci Total Environ 409:218–227. https://doi.org/10.1016/j.scitotenv.2010.08.004

Eberl DD, Smith DB (2009) Mineralogy of soils from two continental-scale transects across the United States and Canada and its relation to soil geochemistry and climate. Appl Geochem 24(8):1394–1404

Egozcue JJ, Pawlowsky-Glahn V, Mateu-Figueras G, Barcelo-Vidal C (2003) Isometric logratio transformations for compositional data analysis. Math Geol 35(3):279–300

Garrett RG (2009) Relative spatial soil geochemical variability along two transects across the United States and Canada. Appl Geochem 24(8):1405–1415

Goldschmidt VM (1937) The principal of distribution of chemical elements in minerals and rocks. The seventh Hugo Muller lecture, delivered before the Chemical Society on March 17th, 1937. J Chem Soc 1937, 665–673. https://doi.org/10.1039/jr9370000655

Grunsky EC (2001) A program for computing rq-mode principal components analysis for S-Plus and R. Comput Geosci 27:229–235

Grunsky EC (2010) The interpretation of geochemical survey data. Geochem Explor Environ, Anal 10(1):27–74

Grunsky EC, de Caritat P, Mueller UA (2017) Using surface regolith geochemistry to map the major crustal blocks of the Australian continent. Gondwana Res 46:227–239. https://doi.org/10.1016/j.gr.2017.02.011

Grunsky EC, McCurdy MW, Perhsson SJ, Peterson TD, Bonham-Carter GF (2012) Predictive geologic mapping and assessing the mineral potential in NTS 65A/B/C, Nunavut, with new regional lake sediment geochemical data. Geological Survey of Canada, Open File 7175, 1 sheet. https://doi.org/10.4095/291920

Grunsky EC, Mueller UA, Corrigan D (2014) A study of the lake sediment geochemistry of the Melville Peninsula using multivariate methods: applications for predictive geological mapping. J Geochem Explor. https://doi.org/10.1016/j.gexplo.2013.07.013

Grunsky EC, Drew LJ, Woodruff LG, Friske PWB, Sutphin DM (2013) Statistical variability of the geochemistry and mineralogy of soils in the maritime provinces of Canada and part of the northeast United States. Geochem Explor Environ Anal 13(2013):249–266. https://doi.org/10. 1144/geochem2012-138

Harris JR, Schetselaar EM, Lynds T, deKemp EA (2008) Remote predictive mapping: a strategy for geological mapping of Canada's North, chapter 2. In: Harris JR (ed) Remote predictive mapping: an aid for Northern mapping. Geological Survey of Canada, Ottawa, Ontario, Canada, pp 5–27 (OpenFile5643)

Harris JR, Grunsky EC (2015) Predictive lithological mapping of Canada's North using random forest classification applied to geophysical and geochemical data. Comput Geosci 80:9–25

Jolliffe IT (2002) Principal component analysis, 2nd edn. Springer, New York, p 487

Mueller UA, Grunsky EC (2016) Multivariate spatial analysis of lake sediment geochemical data. Melville Peninsula, Nunavut, Canada, Applied Geochemistry, https://doi.org/10.1016/j. apgeochem.2016.02.007

Palarea-Albaladejo J, Martín-Fernández JA, Buccianti A (2014) Compositional methods for estimating elemental concentrations below the limit of detection in practice using R. J Geochem Explor. ISSN 0375–6742. https://doi.org/10.1016/j.gexplo.2013.09.003. Accessed 28 Sept 2013

Pawlowsky-Glahn V, Egozcue J-J (2015) Spatial analysis of compositional data: a historical review. J Geochem Explor 164:28–32. https://doi.org/10.1016/j.gexplo.2015.12.010

Pebesma EJ (2004) Multivarilabe geostatistics in S: the gstat package. Comput Geosci 30:683–691

QGIS Development Team (2016) QGIS geographic information system. Open Source Geospatial Foundation Project. http://www.qgis.org/

R Core Team (2013) R: a language and environment for statistical computing. R foundation for statistical computing, Vienna, Austria. http://www.r-project.org

Sayre R, Comer P, Warner H, Cress J (2009) A new map of standardized terrestrial ecosystems of the conterminous United States: U.S. Geological Survey Professional Paper 1768, 17 p. http:// pubs.usgs.gov/pp/1768

Smith DB (2009) Geochemical studies of North American soils: results from the pilot study phase of the North American Soil Geochemical Landscapes Project. Appl Geochem 24(8):1355–1356. https://doi.org/10.1016/j.apgeochem.2009.04.006

Smith DB, Woodruff LG, O'Leary RM, Cannon WF, Garrett RG, Kilburn JE, Goldhaber MB (2009) Pilot studies for the North American Soil Geochemical Landscapes Project—site selection, sampling protocols, analytical methods, and quality control protocols. Appl Geochem 24(8):1357–1368

Smith DB, Cannon WF, Woodruff LG (2011) A National–scale geochemical and mineralogical survey of soils of the conterminous United States. Appl Geochem 26:S250–S255

Smith DB, Cannon WF, Woodruff LG, Rivera FM, Rencz AN, Garrett RG (2012) History and progress of the North American Soil Geochemical Landscapes Project, 2001–2010. Earth Sci Front 19(3):19–32

Smith DB, Cannon WF, Woodruff LG, Solano F, Kilburn JE, Fey DL (2013) Geochemical and mineralogical data for soils of the conterminous United States. U.S. Geological Survey Data Series 801, 19 p. http://pubs.usgs.gov/ds/801/

Smith DB, Cannon WF, Woodruff LG, Solano F, Ellefsen KJ (2014) Geochemical and mineralogical maps for soils of the conterminous United States. U.S. Geological Survey Open-File Report 2014–1082, 386 p. https://pubs.usgs.gov/of/2014/1082/

Venables WN, Ripley BD (2002) Modern applied statistics with S, 4th edn. Springer, Berlin

Woodruff LG, Cannon WF, Eberl DD, Smith DB, Kilburn JE, Horton JD, Garrett RG, Klassen RA (2009) Continental-scale patterns in soil geochemistry and mineralogy: results from two transects across the United States and Canada. Appl Geochem 24(8):1369–1381

Zhou D, Chang T, Davis JC (1983) Dual extraction of R-mode and Q-mode factor solutions. Math Geol 15(5):581–606

Part III
Exploration and Resource Estimation

18

Uncertainty and Probabilistic Risk Analysis

Peter Dowd

Abstract This chapter reviews the general concepts of uncertainty and probabilistic risk analysis with a focus on the sources of epistemic and aleatory uncertainty in natural resource and environmental applications together with examples of
quantifying both types of uncertainty. The initial uncertainty in these applications
arises from the in-situ spatial variability of variables and the relatively sparse data
available to model this variability. Subsequent uncertainty arises from processes
applied either to extract the in-situ variables or to subject them to some form of flow
and/or transport. Various approaches to quantifying the impacts of these uncertainties are reviewed and several practical mining and environmental examples are
given.

18.1 Introduction

This chapter provides an overview of the quantification of uncertainty with a focus
on mineral and energy resources and environmental applications drawing on the
work of the author and his co-authors over the past 30 years. Rarely in mining
applications do initial estimates reconcile with production—there is almost always
some reverse calibration or model revision to achieve an operationally acceptable
agreement. This feedback approach can be a useful means of model calibration but
the production 'reality' is an outcome conditional on the model and data used to
make the production decision and may be biased. The resort to post hoc empirical
calibration is due partly to insufficient data and partly to inadequate accounting for
all sources of uncertainty. This situation will worsen as, increasingly, mineral
resources will be extracted from deeper and/or lower grade deposits, which will
require new technologies and new types of indirect sampling. In applications such
as hydrocarbon extraction, the feedback reconciliation approach is essential because

P. Dowd FREng, FTSE (✉)
The University of Adelaide, Adelaide, Australia
e-mail: peter.dowd@adelaide.edu.au

the in-situ variables can never be directly observed; Caers (2011) gives a comprehensive account of uncertainty quantification for these types of application.

The focus here is on geological applications in which the purpose is to extract material, store material or monitor the flow of fluids or contaminants. In these applications, uncertainty arises from two sources of variability: the in-situ variability of the geology and associated quantitative variables and the variability that is generated by applying processes to the in-situ resource. The basic approach is to combine data with a model to make predictions. Such predictions are meaningless unless accompanied by quantitative measures of the uncertainty of the prediction.

The general focus, particularly in mining applications, has been on the uncertainty arising from sparse data and not on uncertainty arising from the model, even though the model is inferred, and its parameters are estimated, from the sparse data. Variability arising from processes applied to the in-situ resource is either quantified in an overly simplistic manner or is ignored. The additional aspect in these and most spatial applications is that variability (and, therefore, uncertainty) is scale-dependent and may be relevant on multiple scales depending on the application.

18.2   Sources of In-Situ Uncertainty

In the field of uncertainty and probabilistic risk analysis two types of uncertainty are identified: aleatory and epistemic uncertainty (or irreducible and reducible uncertainty). In the generally accepted definitions (e.g., Bedford and Cooke 2001), aleatory uncertainty arises from the inherent variability of a phenomenon and cannot be reduced; epistemic uncertainty arises from incomplete knowledge of the phenomenon and can be reduced by more data, analysis or research. As both types of uncertainty are expressed in terms of probabilities, some authors question the necessity to distinguish between them. Others (e.g. Hora 1996; Winkler 1996) prefer sources of uncertainty rather than types, "the distinction between uncertainties is a matter of choice of scale and is, therefore, mutable." In the geostatistics context, Matheron (1975, 1976, 1978), notes that the empirical basis of uncertainty is the same in both cases and there is no objective criterion to distinguish them. Journel (1994) gives guidelines for modelling uncertainty on which Srivastava (1994) provides critical comment. However, as Winkler (1996) noted "uncertainty is uncertainty but the distinctions are related to very important practical aspects of modelling and obtaining information". This is especially so in the applications given here.

A fundamental difference between geological applications and many others is that each occurrence (orebody, karst system) is unique and, apart from measurement error, once a physical sample is taken at a location and the required variable is measured directly from the sample, there is no longer any uncertainty about the value of the variable at that location. The general geostatistical model includes stationarity, which allows for repeated sampling of the same random variable at different locations. In principle (but not in practice), all locations in an orebody could be sampled and aleatory uncertainty would be eliminated. Thus, in these applications aleatory uncertainty is entirely a function of the amount and quality of data. Epistemic uncertainty arises from the assumed or inferred geological model (e.g., type, or style, of mineralisation). In mining applications, at least in terms of a general model, there may be significant epistemic uncertainty during early stages of proving a deposit when geological models are inferred from sparse data. Model uncertainty may persist in later stages in terms of the specific characteristics or parameters of the model.

In some natural resource applications, the variables that define the resource can never be directly observed. For example, in hot dry rock (HDR) enhanced geothermal systems, the variable of interest is the combination of natural and stimulated fractures that form connected networks to extract heat. These fractures, at depths of up to 4.5 km, can never be directly observed or measured; their locations, extents and characteristics can only be inferred from micro-seismic events generated by fracture movement, stimulation and propagation (e.g., Xu and Dowd 2014). In these applications, the detailed model can never be known irrespective of the amount of data available. As mineral resources are extracted from increasingly deeper deposits there will be a move from physical samples, from which variables are directly measured, to sensed proxy variables and a move from traditional mining methods to in-situ recovery. For indirectly sensed variables, the aleatory uncertainty of the required variable (e.g., porosity) is largely due to the quality of the relationship with the directly sensed proxy variable (e.g., acoustic impedance), which could be classified as measurement, or interpretation, error.

Thus, although both sources of in-situ uncertainty in these applications are functions of the amount of data, it is useful to distinguish between them in quantifying uncertainty. Hereafter, epistemic uncertainty is used to mean conceptual or descriptive geological models as well as quantitative parametric models that describe spatial variability and in which parameter values are calculated or inferred from data.

Although epistemic uncertainty is recognised, it is largely ignored in practice. Once a model is assumed or inferred and/or its parameters are inferred or estimated from the available data, all measures of uncertainty are based on the data; in most applications, the model of spatial variability is implicitly assumed to be known with certainty. In other fields, there has been a longstanding recognition of the importance of identifying and quantifying both sources of uncertainty and of propagating them into a complete systems model (e.g., Bedford and Cooke 2001; Helton et al. 2004; Oberkampf et al. 2002, 2004). In natural resource applications, particularly mining, the emphasis has largely been on aleatory uncertainty with implicit acceptance that epistemic uncertainty is negligible. Geostatistical simulation is widely used to quantify the effects of limited data on resource modelling and estimation (aleatory uncertainty) but the model (e.g., variogram, spatial pattern) is generally assumed to be perfectly known (no, or negligible, epistemic uncertainty).

18.3   Transfer Uncertainty

A further complication in mineral and energy resources is that there are additional significant sources of uncertainty in extraction and processing to produce a final product. To borrow a petroleum industry term these might be called transfer, or process, functions and the associated uncertainties, transfer or process uncertainty. A general approach to integrating this source of uncertainty is to quantify all sources of in-situ uncertainties and propagate them into simulated transfer processes (e.g., blasting, selective loading, transport, mineral processing).

In resource extraction applications, it is useful to distinguish two broad types of process (or transfer) uncertainty:

(1) The uncertainty associated with in-situ variables that is propagated into processes applied to them. This might be termed passive in the sense that it does not change spatial variability. An example is the impact of grade uncertainty on mine design, which could be assessed by applying the same design process (e.g., optimal open-pit) to a range of simulated realisations of grades.
(2) The uncertainty transferred, or propagated, to in-situ variables by applying processes to them. This might be termed active as the process changes spatial variability. Changes in spatial variability can be predicted by modelling the process. An example is blasting a block of ground from which ore is selected.

18.4   Consequences of In-Situ Uncertainty

There are broadly two aspects of a geological model used in mineral resource applications: the generic type (e.g., stratiform silver/lead/zinc orebody) and the unique aspects that distinguish a specific orebody within the type (e.g., faulting, folding, degree of spatial continuity and of regularity of orebody boundaries). In general, for mineral deposits the first of these is known with near certainty at a relatively early stage but the distinguishing aspects and the relevant scales on which these aspects occur may not be known until much later. In these applications, the two types of in-situ uncertainty are not independent. The sampling scale (e.g., drilling grid) is determined, or at least significantly informed by, the geological model; the sampling scale determines the data, the spatial variability of which is the aleatory uncertainty; the parameters of the model are estimated by the data.

The Stekenjokk mine in Sweden provides a striking example of the consequences of epistemic uncertainty. Boliden Mineral AB mined this massive copper-zinc-silver orebody from 1976 to 1988 and processed a total of 8 M tonnes of ore. Prior to mine development the drilling grid was 20 m × 20 m and, in places, 20 m × 10 m. Figure 18.1 is an idealised, but typical, vertical cross-section through the orebody showing the drill-hole intersections with the ore. Drilling data were combined with the assumed geological model to generate the estimated orebody boundaries.

Figure 18.1 shows the complex, multi-directional folding of ore zones encountered in mining. The practical consequences of these predictions were significant (Hoppe 1978):

- Inappropriate choice of mining methods and mining equipment.
- Increased ore dilution, mining costs, development and processing provisions.
- Complications of highly mechanised equipment purchased for a simpler mine.

In principle, the problem could have been resolved by more appropriate sampling but the "appropriateness" of sampling was determined by the assumed geological model. In addition, sampling is constrained by cost (relative to the value of the mined product) and the cost of a drilling grid capable of capturing the folding may well have been prohibitive.

Geological models are only as good as the quality and interpretation of the data and the appropriateness of the scale on which the data are collected. Stekenjokk is an extreme (but not unique) example of epistemic uncertainty that could only be

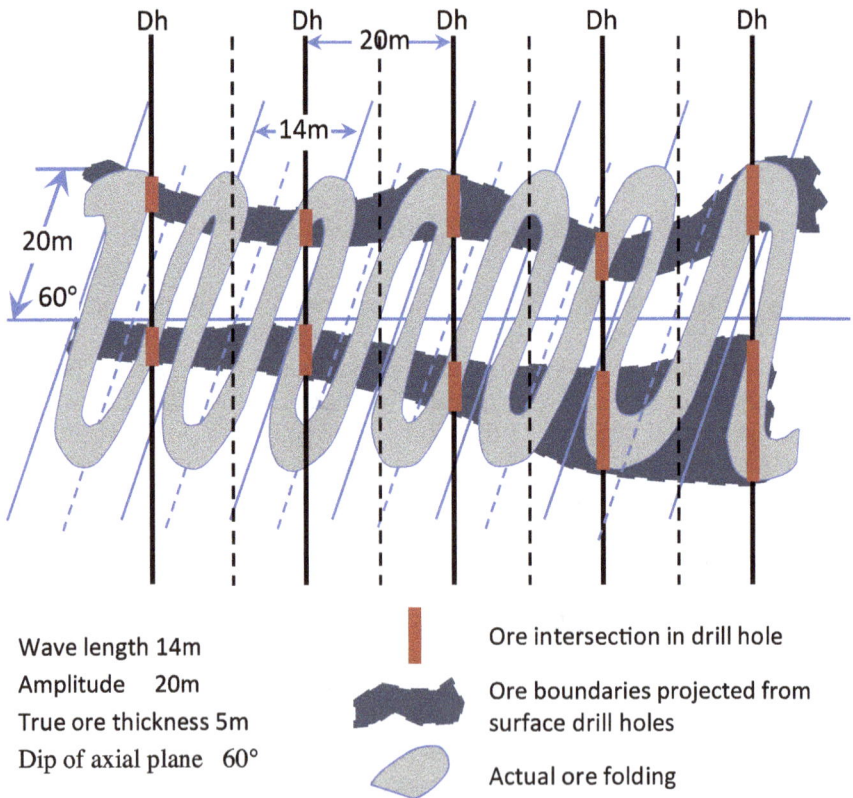

Wave length 14m
Amplitude   20m
True ore thickness 5m
Dip of axial plane 60°

Ore intersection in drill hole

Ore boundaries projected from surface drill holes

Actual ore folding

Fig. 18.1 Interpolation of ore continuity from surface drilling data prior to mine development; adapted from Hoppe (1978)

reduced to an acceptable level by more data. However, this observation is some-what circular: the geological model depends on the amount of data/information available but the data type and collection are informed by the assumed model.

18.4.1   Scale and Variability Example: Hilton Orebodies Australia

This example is from a study of a complex group of three silver/lead/zinc orebodies at what, at the time, was known as the Hilton mine in north-western Queensland, Australia. The full study is given in Dowd and Scott (1984) with a later study in Dowd et al. (1989).

The Hilton orebodies are 22 km north of Mt Isa, one of the world's largest stratiform base metal deposits. The Hilton orebodies have a similar diagenesis to the Mt Isa orebodies with mineralisation occurring in the same dolomitic shale. The study was undertaken at the pre-feasibility stage and all original drilling, sampling and interpretation were influenced by 50 year's mining experience at Mt Isa. Although the Mt Isa and Hilton styles of mineralisation are similar, the Hilton orebodies are structurally more complex and less continuous.

Two test areas were extensively drilled to provide detailed information for a geostatistical study to determine optimal drilling densities for mine planning pur-poses. The holes were drilled from access drives as fans on cross-sections spaced 10 and 20 m apart. One such cross-section is shown in Fig. 18.2 in which the holes intersect the main 2 orebody footwall lens (2 O/B FW) at approximately 5 m centres. The dark blue outlines in Fig. 18.2 are the orebody boundaries estimated from the drill-hole data on the cross-section and on the cross-sections on either side. In the feasibility stage cost would prohibit such a drilling density over the entire orebody. Given the density of the drilling these estimated boundaries could be regarded as reality on all practical scales.

The effects of other drilling densities were assessed by removing drill data to create new datasets; e.g., removing every second drill-hole on a cross-section yields a 10 m spacing. Datasets for 5, 10, 20 and 40 m drill spacing were used in the study. Orebody boundaries were estimated for each drilling density and the results were given to mining engineers to design stopes. As an example, the estimated orebody boundaries for 20 m drill spacing is shown in Fig. 18.3. As expected, these boundaries are much smoother (less variable, more continuous) than the "reality" represented by the boundaries estimated from the 5 m spacing dataset. The vari-ability of the boundaries is critical in the choice of mining method: the variability of boundaries and their exact delineation are less critical if a bulk mining method is adopted than if more selective methods are used. The original mining method was cut and fill followed later by sub-level open stoping and bench mining.

Figure 18.4 shows the 5 m interpolation overlaid on the 20 m interpolation. Taking the 5 m interpolated boundaries as reality, all visible light blue areas

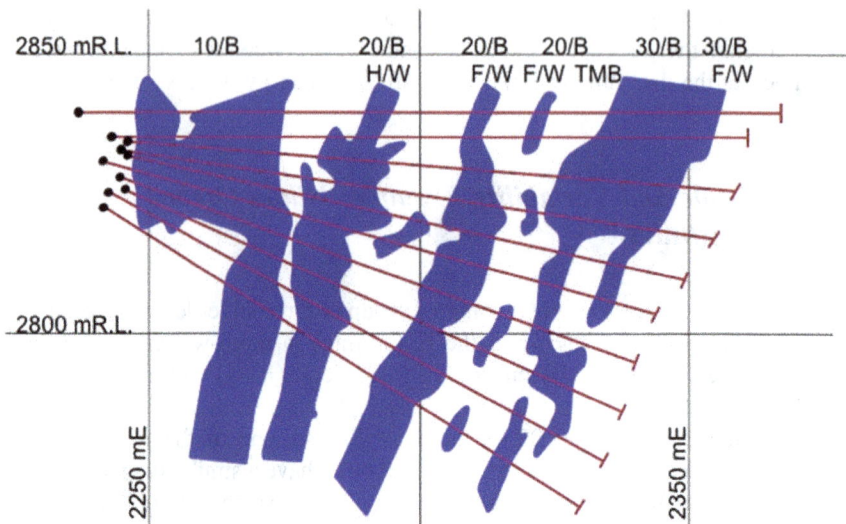

Fig. 18.2 Cross-sectional interpretation based on 5 m drill spacing

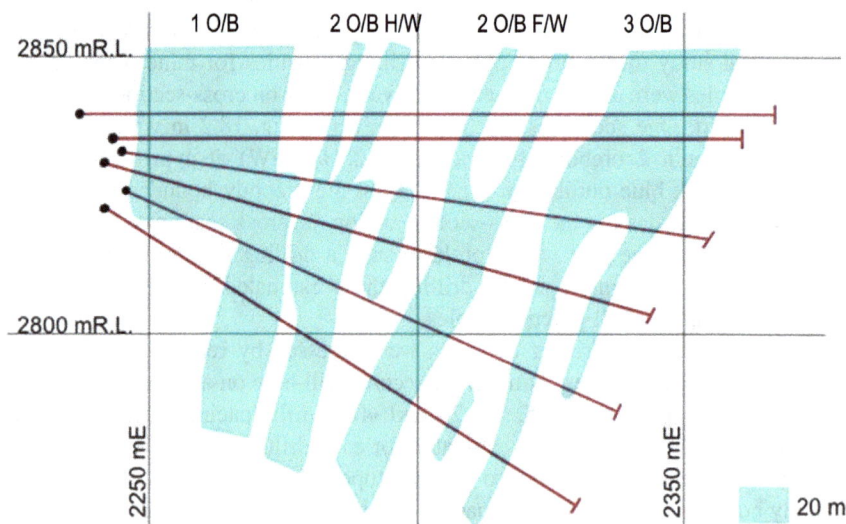

Fig. 18.3 Cross-sectional interpretation based on 20 m drill spacing

represent ore dilution arising from planning and extraction based on the 20 m interpolated boundaries.

Figure 18.5 shows the 20 m interpolation overlaid on the 5 m interpolation. Again, taking the 5 m boundaries as reality, all visible dark blue areas represent the ore loss arising from planning and extraction based on the 20 m interpolated

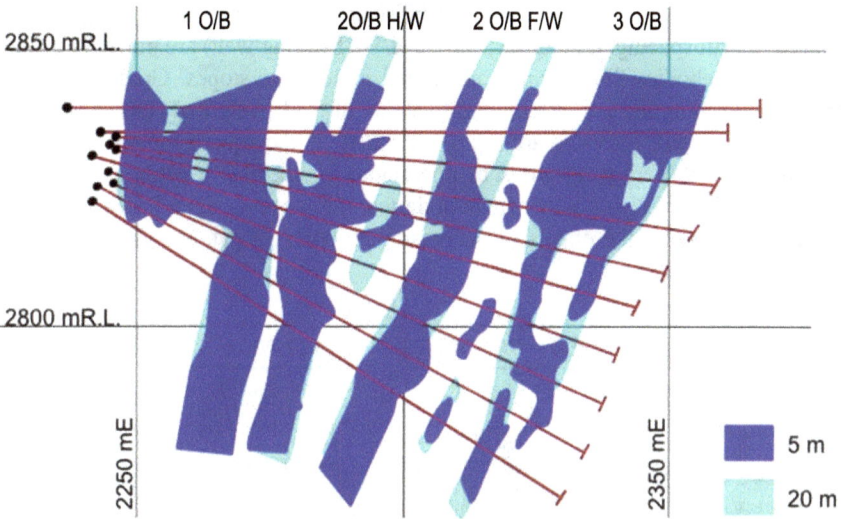

Fig. 18.4 Overlay of 5 m interpolation on 20 m interpolation

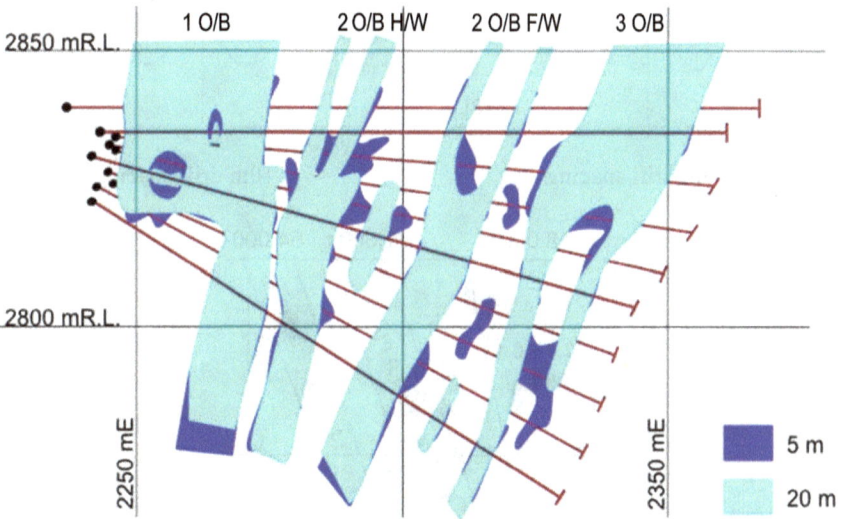

Fig. 18.5 Overlay of 20 m interpolation on 5 m interpolation. Based on 20 m model, all visible dark blue areas represent ore loss

boundaries. Of course, the perfect selection and the adherence to estimated boundaries during production implied by this exercise are not entirely realistic. However, the impact on the choice of mining method, on the predicted grades and tonnages, and on economic outcomes is real.

The outputs from the stope design exercise are summarised in Fig. 18.6 for 5, 10 and 20 m drill spacing. Orebodies 1 and 2 H/W (hanging wall) are mined in a single stope and orebodies 2F/W and 3 are mined in separate stopes. Grades were estimated by kriging and are in metal equivalents of lead (weighted sum of lead, zinc and silver grades); intervals are $\pm 2\sigma_K$ where σ_K is the square root of the kriging variance and is used as an index of uncertainty rather than a confidence interval. Taking the 5 m designs as actual boundaries, the stope designs based on 10 and 20 m drilling show the effects of decreasing amounts of data on planned tonnage and average grade.

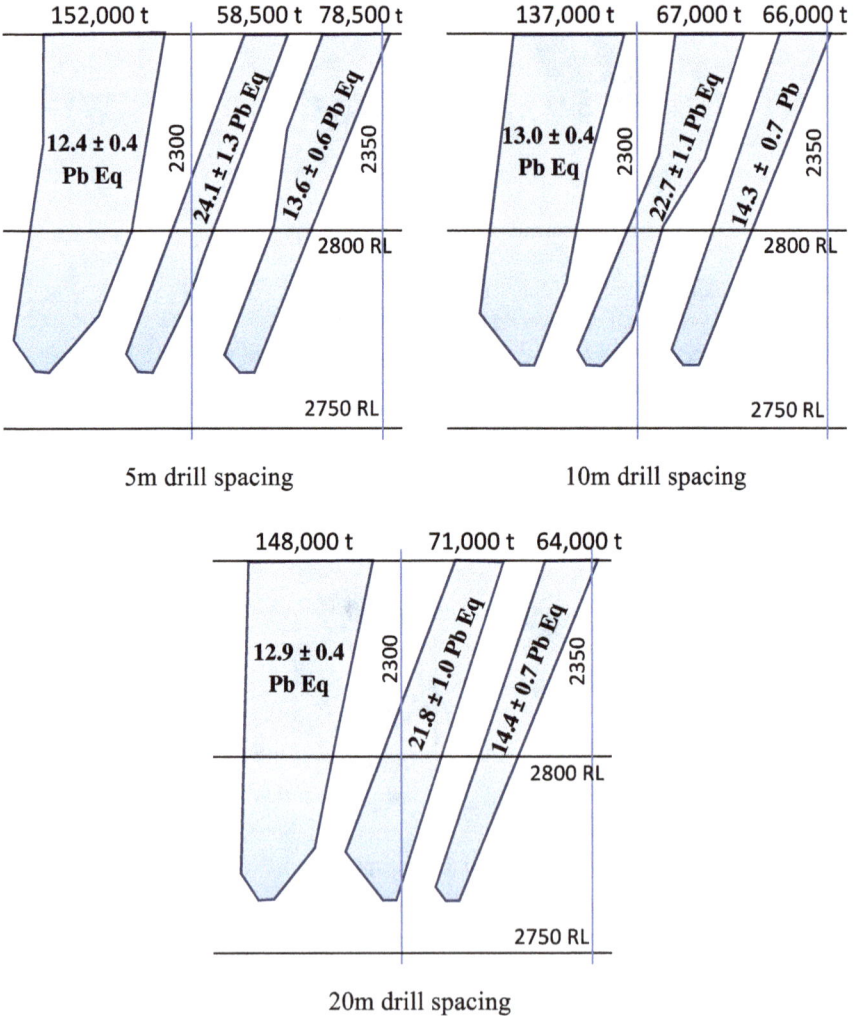

Fig. 18.6 Stope designs with contained tonnages and grades for 5, 10 and 20 m drill spacing for orebodies 1 and 2 HW (left); 2 FW (centre) and 3 (right)

Table 18.1 Differences in tonnes and grades of stopes compared with 5 m designs

Stope	Drill spacing (m)	Change in ore tonnes (%)	Change in grade (%)	Change in metal tonnes (%)
1	10	−10.2	+4.2	−6.4
	20	−3.0	+3.6	−0.5
2	10	+14.5	−5.9	+7.8
	20	+21.4	−9.6	+9.7
3	10	−15.9	+5.0	−11.8
	20	−18.5	+5.5	−14.0

The stope designs are based on the data and interpretations from the respective drilling densities but the grades and tonnages are estimated using all data (5 m drill spacing). Assuming the data from the 5 m drill spacing gives the closest possible quantification of reality on all practical scales then the grade and tonnage of the 10 and 20 m stope designs estimated from all data can be regarded as sufficiently close to the real tonnage and grade that could be recovered from the designs.

The effects of data density on grades and tonnages are summarised in Table 18.1. As an example, using the 20 m drill spacing data to design stope 2 (the high-grade orebody 2 footwall) would increase tonnage by 21.4% and reduce grade by 9.6%. There would be an increase in metal tonnage of 9.7% but this would at the cost of mining, hauling and processing the additional ore tonnage.

Whilst the effects of data on a specific type of mining are of interest, the more important issue is the effect of the assumed geological model on the choice of mining method. The initial geological model was influenced by the knowledge accumulated over a long period of mining in the neighbouring Mt Isa orebodies. The detailed analysis described here enabled the effects of the greater complexity and less continuity of the Hilton orebodies to be systematically quantified, thereby significantly reducing the impact of epistemic uncertainty and contributing to the selection of the most appropriate mining method and mine design.

18.5   Quantifying Epistemic Uncertainty

In the Hilton example, geological model uncertainty was addressed at the significant cost of more samples—effectively eliminating the epistemic uncertainty on the operational scale through more data and analysis. With the hindsight of the additional data and analysis, and on the assumption that the test volume is sufficiently representative of the remainder of the orebodies, the epistemic uncertainty associated with various drilling grids could be quantified. This would allow assessment of the value of additional information against the cost of collecting it and/or the operational cost of not collecting it. Stekenjokk is an example of the practical consequences of proceeding with an unacceptable level of epistemic uncertainty.

There is an extensive literature on using Bayesian probability to quantify epistemic uncertainty particularly to combine sources of uncertainty (e.g., Winkler 1981; Sankararaman and Mahadevan 2011) and to incorporate expert knowledge and informed guesses in the form of subjective probabilities. It can be argued that subjective probabilities are used implicitly throughout geostatistical analysis, modelling, estimation and simulation irrespective of the amount of data. Expert knowledge/judgment guides variogram calculation and interpretation, choice of training images, domaining, sample differentiation, choice of estimation or simulation method and validity of outputs. There is, however, a distinction between the explicit subjective probability of informed guesses and possible geological models and the implicit subjectivity in inferring model parameters from quantitative data.

In the remainder of this chapter, a distinction is made between model uncertainty and uncertainty of the parameters of a specific model. Many authors do this although in some cases the former may be a case of the latter e.g., it might be argued (with some difficulty) that Stekenjokk was a matter of incorrect structural parameters (degree of folding). A more convincing argument could be made for the Hilton case—the initial assumed model was a Mt Isa type stratiform orebody and the final agreed version was a more complex and less continuous version of the latter.

In addition to Bayesian approaches, others include evidence theory: Shafer (1976) and Dempster (1968); fuzzy sets: (Zadeh 1965); and possibility theory: Zadeh (1978) and Dubois and Prade (2001). These and other approaches are extensively used to quantify uncertainty in risk analysis and a good coverage of probabilistic risk analysis is given in Bedford and Cooke (2001).

Over the past 30 years, all these approaches have been used to incorporate model uncertainty in geostatistical estimation and simulation and the following list is intended as representative rather than exhaustive. Omre (1987) used Bayesian kriging to include qualified guesses when few data are available; the weight assigned to the guess increases as the amount of data decreases.

Fuzzy kriging has been proposed as a means of including aleatory uncertainty (in the sense of inaccurate or imprecise measurements) and epistemic uncertainty (imprecise variogram parameters) in estimation. Uncertain data will, of course, lead to an uncertain variogram but certain (accurate, error-free) data will not necessarily lead to a certain variogram. Diamond (1989) proposed fuzzy kriging to deal with uncertain or imprecise data. Bardossy et al. (1988, 1990a, b) proposed fuzzy kriging for dealing with both sources of uncertainty but the computational cost hindered its use. More recently, Loquin and Dubois (2010a, b) have developed these approaches in computationally feasible forms. Bandemar and Gebhardt (2000) combine fuzzy kriging with Bayesian incorporation of prior knowledge. Bardossy and Fodor (2004) provide a comprehensive coverage of the use fuzzy set theory to quantify geological uncertainty and consequent risk.

Srivastava (2005) used probabilistic modelling of ore lenses to account for uncertainty in the boundaries of geological domains that constrain grade occurrence. Dowd (1986, 1994) and Dowd et al. (1989) used deterministic and probabilistic methods for the same purpose in estimating and simulating grades.

Verly et al. (2008) quantified geological model uncertainty in a porphyry copper deposit by simulating the four principal characteristics of porphyry models: faults defining fault blocks; faulted rock types within fault blocks; un-faulted intrusive and breccia bodies and alteration and copper grade shells.

Maximum likelihood estimation of spatial model parameters has been widely reported in geostatistical applications: Mardia and Marshall (1984), Kitanidis and Lane (1985), Zimmerman (1989), Dietrich and Osborne (1991) among others. Pardo-Igúzquiza and Dowd (1997a, b, c, 2003, 2013), Dowd and Pardo-Igúzquiza (2002) and Pardo-Igúzquiza et al. (2013) used maximum likelihood estimates of variogram parameters and associated uncertainties to incorporate the effects of model uncertainty in simulation and estimation.

For categorical variables, such as geological shapes and surfaces, multiple point statistics simulation provides a means of specifying possible geological scenarios in the form of alternative training images. Caers (2011) uses different training images to introduce geological model uncertainty into the simulation of oil reservoirs. Park et al. (2013) use history matching to quantify the uncertainty of facies models in the form of alternative training images. Hermans et al. (2014) choose among several geological scenarios in the form of possible training images using geophysical data and Bayes rule to compute the conditional probabilities of the alternative training images given the geophysical data.

With a few notable exceptions, in most mining applications the geological (model) uncertainty from the feasibility stage onwards can be limited to uncertainty in model parameters rather than uncertainty about the general model (e.g., stratiform, vein, disseminated). However, for cases where fundamental (and a priori, unverifiable) assumptions are/must be made about the general model, as in oil and gas applications or applications in which physical processes give rise to the variables (e.g., HDR fracture occurrence and propagation), it is essential to test the sensitivity of these assumptions by reconciling the consistency of outputs (e.g., heat production from a geothermal reservoir) with predicted responses to inputs (e.g., fluid flow through fracture networks). The fundamental difference between these cases and mining applications is that ultimately the latter can be directly observed.

On the assumption that the most important characteristics of the underlying model can be captured in several parameters of a broad model, the uncertainty in the parameter estimates can be quantified by generating a set of parameter values using an appropriate set of rules; simulating the spatial random variable(s) using these parameter values; and repeating this process a sufficiently large number of times. Methods for sampling parameter values include Maximum Likelihood, Bootstrap methods (Olea et al. 2015), Bayesian analysis (Kitanidis 1986) and, in multiple point statistics simulations, Bayesian selection of alternative templates or training images (Park et al. 2013; Hermans et al. 2014) and clustering combined with system responses (Caers 2011).

The following two examples illustrate the use of maximum likelihood in model selection and parameter inference and the propagation of the associated uncertainties into geostatistical simulation for environmental and mining applications.

18.5.1 Example: Transmissivity Uncertainty

This example is taken from Dowd and Pardo-Igúzquiza (2002). The data are from Gotway (1994) and comprise 41 transmissivity measurements in the Culebra Dolomite formation in New Mexico. The original application was for nuclear waste site assessment, where uncertainty in the groundwater travel time of a particle is assessed through its probability density function, which is estimated by running groundwater flow and transport programs with different transmissivity field inputs. These inputs are generated by conditional simulations of transmissivity.

The data are the logarithms of transmissivity in $m^2 \, s^{-1}$ and the data locations are shown in Fig. 18.7 together with a histogram of the log-transmissivity data.

Maximum Likelihood was used to estimate the parameters of an exponential covariance model of the residuals for drift orders 0, 1 and 2. Although drift is a deterministic component of the universal model, in practice the coefficients are estimated from the available data and are thus random variables with the means and standard errors given in Table 18.2 for the optimal (determined by the Akaike information criterion) drift model of order 1: drift $(x, y) = \beta_0 + \beta_1 x + \beta_2 y$. The estimated covariance parameters for $k = 1$ are given in Table 18.3 and the variogram is shown in Fig. 18.8.

In this case, as there is no nugget variance, the range and sill are estimated independently. The correlation between range and sill is thus zero and any combination of values of the two parameters inside their respective intervals is inside the 95% confidence region as shown in Fig. 18.9a. The drift coefficients are also independent of the sill and the range. As the estimated drift coefficients are correlated, not every combination of the three parameter values is equally reliable, i.e. values inside the 95% confidence interval of the parameters taken together may not be inside the 95% confidence interval for each individual parameter. The confidence interval is an ellipsoid. Figure 18.9b shows the 95% confidence region for (β_1, β_2) when the third coefficient the model is set to the estimated value given in Table 18.3.

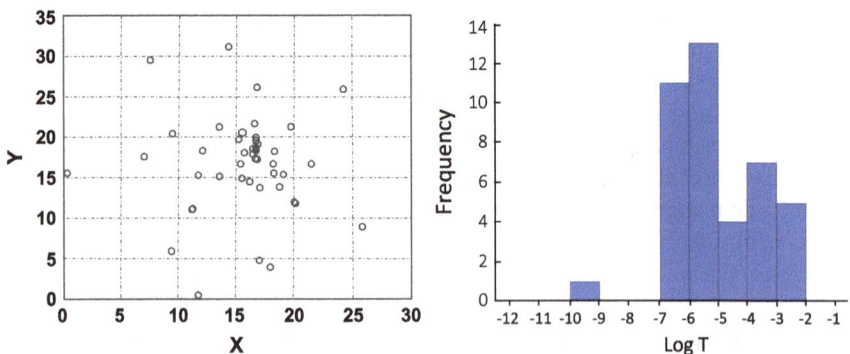

Fig. 18.7 Data locations (distances in km) and histogram of log transmissivity data

Table 18.2 Maximum likelihood estimates of drift coefficients

Parameter	Estimate	Stand. error
β_0	−1.6062	0.8653
β_1	−0.2245	0.0426
β_2	−0.0141	0.0323

Table 18.3 ML estimates of range and sill: exponential covariance

Sill	Stand. error	Range	Stand. error
1.28	0.284	1.99	0.667

Fig. 18.8 Semi-variogram of the residuals for $k = 1$ and maximum likelihood model fitted: sill 1.28, range 1.99 km (effective range ~6 km)

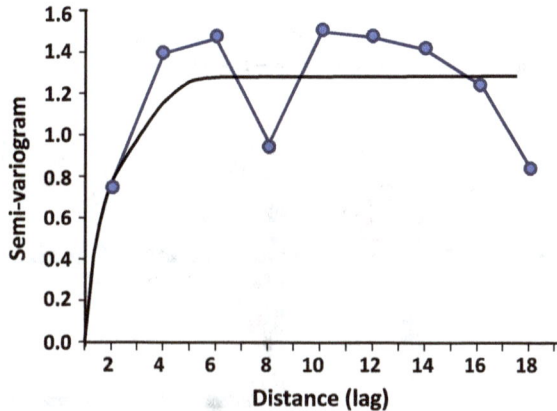

The effects of model uncertainty on simulation outputs are illustrated by generating six simulations for each pair of values A, B, C, D and E in Fig. 18.9; each set of simulations was started with the same random number seed. The simulations are shown in Fig. 18.10. The differences between corresponding simulations (e.g., first simulation in each of A, B, C, D and E) for the five sets of parameters reflect the model uncertainty, which could be quantified further by simulating flow and transport through the simulated transmissivity realisations.

18.5.2 Example: Coal Resource Risk Assessment

One of the most significant contributors to the total risk in the evaluation of coal-mining projects is the uncertainty of the resource tonnage and quality characteristics, often called the resource risk. This example is from the As Pontes deposit in Galicia, Spain (Pardo-Igúzquiza et al. 2013). The most significant variable in the assessment of resource uncertainty is the thickness of the coal seam. Figure 18.11 shows the data locations at which seam thickness is measured together with the estimated variogram values and the manually fitted (isotropic) variogram model.

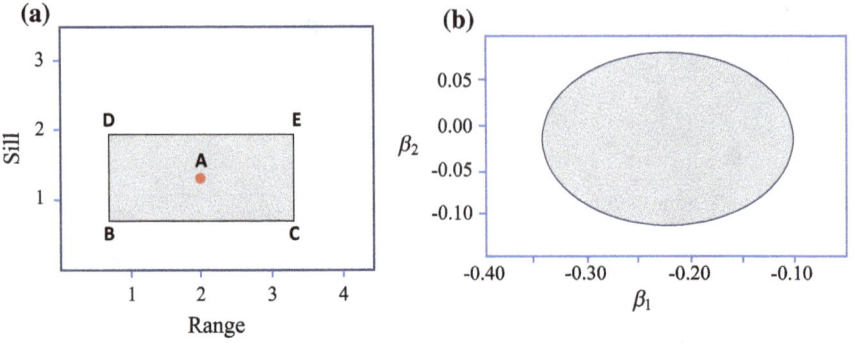

Fig. 18.9 a (left) 95% confidence region for sill and range; b (right) confidence region for drift parameters β and β with β $= -1.6062$

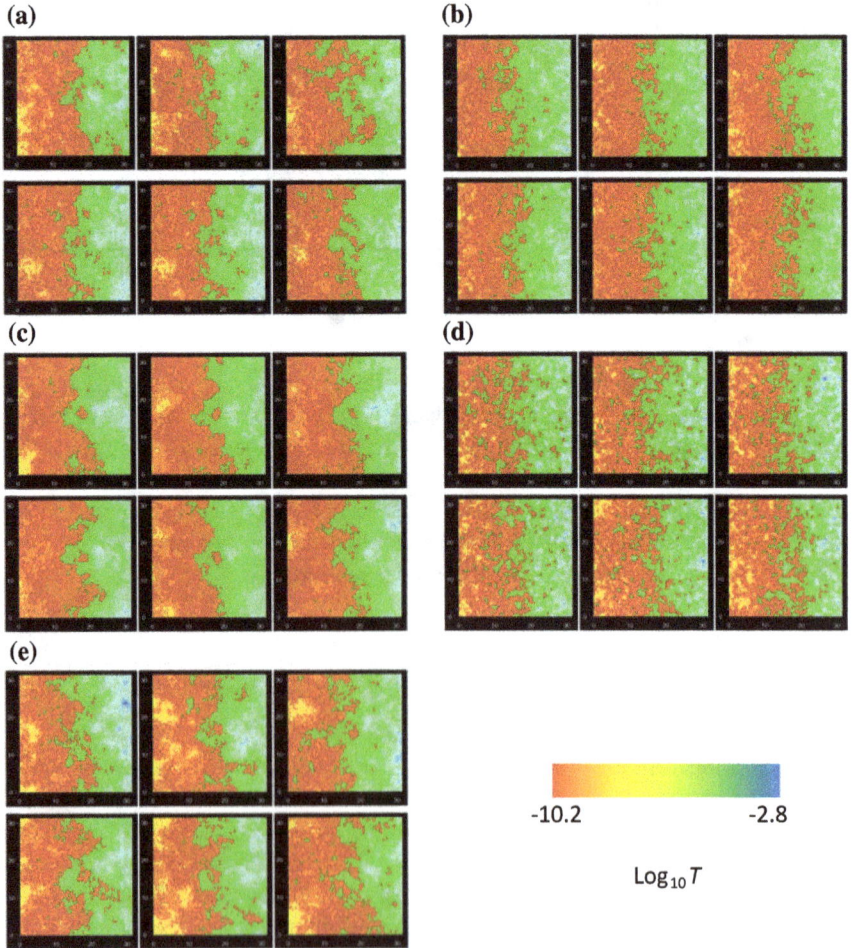

Fig. 18.10 Outputs from six simulations using the variance and range parameters denoted by the mean values A and the extreme values B, C, D and E in Fig. 18.9

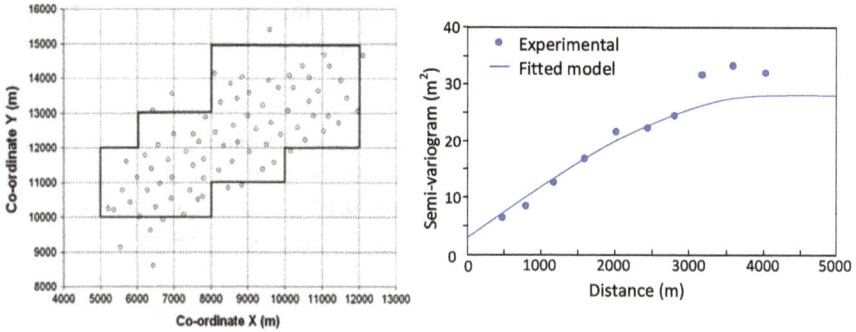

Fig. 18.11 (Left) drill-hole locations and boundary of the study area. (Right) Variogram and manually fitted model for seam thickness

Spherical model variograms for seam thickness:

- Manual fitting: $a = 4128$ m, $C_0 = 3$ m^2 and $C = 25$ m^2.
- Maximum Likelihood: $a = 4460$ m, $C_0 = 4$ m^2 and $C = 23$ m^2.

Although the maximum likelihood estimates of the parameters are very similar to those estimated by visual fitting, maximum likelihood has the advantage of providing estimates of the uncertainty of the parameters. For illustrative purposes, resources were computed as tonnage from panels with thickness above a threshold defined by the 25th percentile of the sample data and equal to a thickness of 8.65 m. The kriged resource volume is 1.97×10^8 m^3.

Sequential Gaussian simulation was used to generate realisations of the thickness of the seam. To quantify the uncertainty in the estimated resource, a total of 870 simulations were generated using the 'certain' variogram (maximum likelihood

Fig. 18.12 Conditionally simulated realisation of coal seam thickness

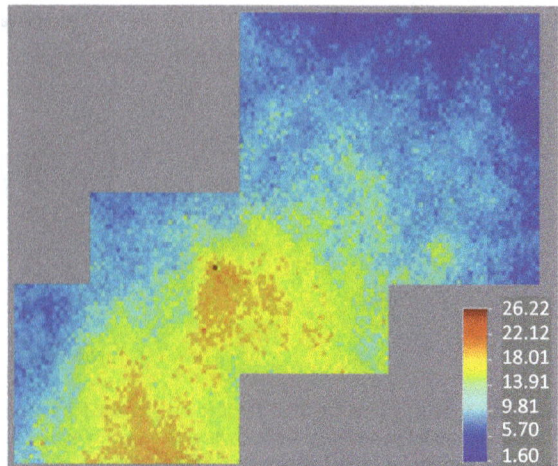

parameters) and the total resource was calculated for each simulation. The histogram of the 870 simulated resources quantifies the uncertainty of the estimated resources. An example simulation is shown in Fig. 18.12.

The parameter space $\{r_0, a, \sigma^2\}$ comprising respectively the nugget/variance ratio, range and variance, is used to quantify the uncertainty in the model. The parameter values were divided into discrete steps of 0.05 for r_0 in the interval [0, 1]; 700 m for a in the interval [1,000, 15,000] and 0.1 for σ^2 in the interval [0.6, 2.6]. There are 268 models of triplets $\{r_0, a, \sigma^2\}$ that lie inside the 75% confidence region. As these models are not equally probable, the probabilities are normalised so that they sum to 1.0 and each model is included as many times as indicated by its normalised probability (i.e., probability sampling in which, for example, a model with a normalised probability of 0.35 comprises 35% of the total simulated triplets). A total of 870 simulations were used.

Histograms of the total resources for the 870 simulations, with and without the uncertainty of the variogram model parameters, are given in Fig. 18.13. There is no significant difference in mean resource values for the certain and uncertain values.

The 95% confidence interval for the total resource assuming the variogram is known with certainty is $[1.88 \times 10^8, 2.19 \times 10^8]$ m^3 and $[1.90 \times 10^8, 2.23 \times 10^8]$ m^3, when the uncertainty of the variogram model is included. The latter is slightly higher than the same interval calculated under the assumption that the variogram is known with certainty. However, the probability that the total resource will be greater than 2.0×10^8 m^3, is 0.59 when the uncertainty of the variogram parameters is ignored and 0.75 when the uncertainty of the variogram parameters is propagated into the simulated realisations. In other words, whilst there is no significant difference in the mean resource for the two sets of simulations, the difference in the two distributions (because of different variances) is sufficient to generate significantly different resource estimates above selected cut-offs.

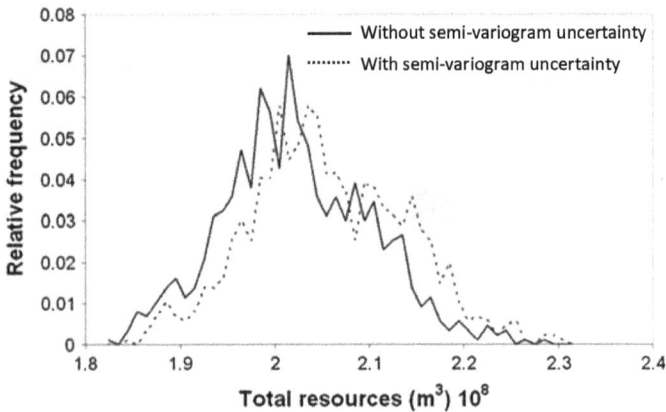

Fig. 18.13 Histograms of total resources calculated by geostatistical simulation assuming the variogram model parameters are known with certainty (solid line) and including the uncertainty of the semi-variogram model parameters (dashed line)

In this case, the differences in the total volume of resources, with and without quantification of semi-variogram uncertainty, are small but the consequence of selecting from the distribution of possible resources is significant. This illustrates a general principle: the estimated total resource and the mean simulated resource, with and without semi-variogram uncertainty, may not differ significantly but the distributions of the two simulations will differ because of the different variances. Similarly, selecting panel values above a threshold from the set of estimated panel thicknesses or from a set of simulated panel thicknesses will yield different results.

In general, the outcome from the simulations with and without semi-variogram uncertainty depends on the deposit and the amount of data available. Evaluation of model uncertainty is critical in resource risk assessment even if it is ultimately found that there is no practical difference between resource estimates obtained by ignoring or including semi-variogram uncertainty. This example also has important implications for compliance with resource and reserve reporting codes, most of which use terms such as, or equivalent to, *the amount of error* [associated with an estimate], *the level of accuracy* [of an estimate], the *level of confidence* [in a reserve statement], and *levels of geological confidence* (words in italics are quoted from JORC 2012). Whilst all reporting codes currently use these terms qualitatively they all have specific quantitative meanings in statistics, probability and risk assessment and are increasingly being referred to explicitly in reporting codes.

18.6   Quantifying the Effects of Transfer Uncertainty

An example of passive transfer uncertainty is the variation in open-pit size and shape as a function of grade uncertainty as shown in Fig. 18.14 taken from a study of a small gold orebody (Dowd 1995, 1997). The impacts of these types of uncertainty can be quantified by standard applications of geostatistical simulation. Dimitrakopoulos and co-workers have made significant contributions to the integration of in-situ grade and geological uncertainty into optimization algorithms (e.g., Dimitrakopoulos et al. 2002; Goodfellow and Dimitrakopoulos 2013).

More challenging is the impact of propagating in-situ uncertainty through the mining (extraction) process. The critical component of most metalliferous open-pit mining operations is ore selection, i.e. the minimisation of ore loss and ore dilution during extraction. In general, extraction comprises drilling, blasting and loading, all of which are planned and designed on uncertain models of local geology and grade. The conversion of the in-situ block model resource to a realistically recoverable reserve may, in many instances, be the most significant source of uncertainty in reserve estimation. The usual assessment of recoverable reserves, for example, is limited to a simple volumetric exercise in which ore recovery is assessed as a function of applying a range of selection volumes to a simulated orebody or an even simpler volume-based adjustment of the variance of estimated block values. These simplistic approaches ignore the practicalities of the mining, selection and loading processes—blast design, behaviour and performance; equipment type, size and

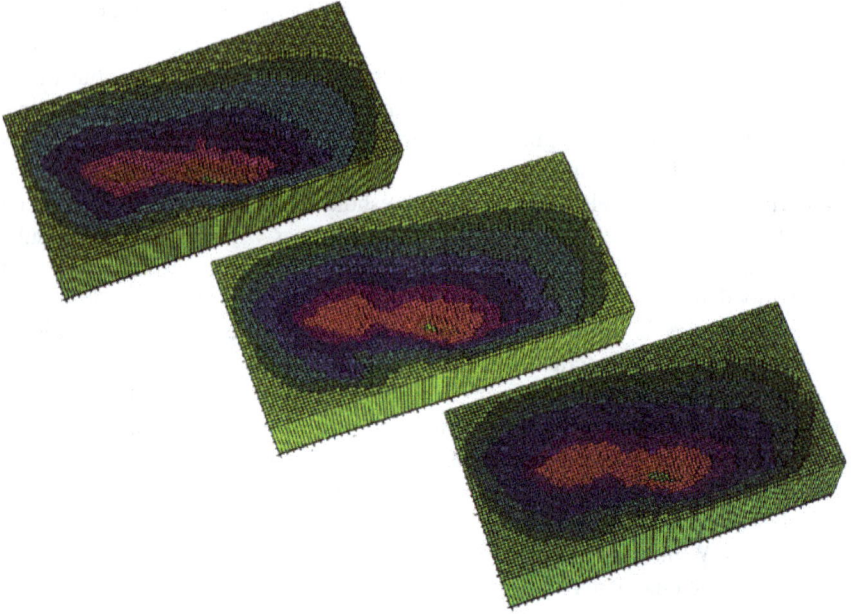

Fig. 18.14 Optimal open pits generated from 100 simulations of a small gold orebody. Top: maximum volume; centre: median volume; bottom: minimum volume

operation; ore displacement during blasting and loading; and ability to identify ore zones within a blast muck pile. In many applications, the uncertainties introduced by these technical processes are at least as significant as those that derive from the in-situ spatial characteristics of grades and geology.

An approach to quantifying transfer process uncertainty for blasting and loading comprises:

- generation of an in-situ model of the orebody comprising the grade, geology, geomechanical properties and grade control variables within small volumes determined by the smallest selectable volume within a blast muck-pile;
- definition of a blast volume comprising a large number of in-situ model volumes, and subjecting it to a blast simulator, which effectively moves each component model volume to its final resting place in the blast muck-pile; and
- application of simulated selective loading processes to the simulated blast muck-pile to determine the selectivity that can be achieved by various sizes of loader and types of loading and to quantify ore dilution and ore loss.

The in-situ model, representing perfect knowledge at all relevant scales, is obtained by geostatistical simulation. An in-situ model that represents the reality of knowing only the data and information that are available from specific grade control drilling and sampling grids can be obtained by sampling the geostatistically simulated model on a specified grid. The volumes comprising the in-situ model are then populated by estimates based only on the data corresponding to the specified grade-control drilling and sampling grids. Different drilling and sampling grids can be used to generate different models, each reflecting the levels of data and information available. Selectivity can then be assessed as a function of the drilling and sampling grids as well as the size and type of loader. Performance is assessed against the ideal selectivity that can be achieved on the perfect knowledge model, comprising the simulated values of each component volume. Applying costs, prices and financial criteria enables an optimal selection of the grade control drilling grid, size of loader, type of loading and even blast design.

The following case study (Dowd and Dare-Bryan 2004) is based on the Minas de Rio Tinto SAL open-pit copper mine at Rio Tinto, southern Spain, which is typical of a low-grade operation in the later stages of its life. Ore/waste delineation for selective mining is difficult because the head grades are near the economic cut-off grade and there are no clear geological controls on the mineralisation.

Sequential Gaussian simulation, with the blast-hole grades as conditioning data, was used to generate realisations of each mining bench on a block grid of 0.5 m × 0.5 m × 0.5 m, the grid determined based on blast and selection criteria.

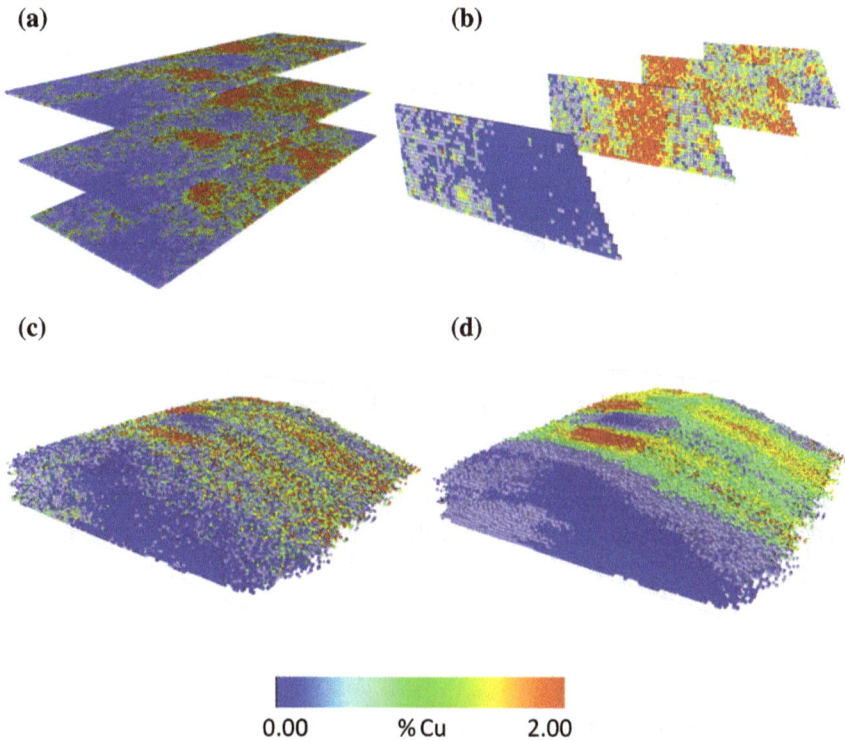

Fig. 18.15 **a** simulated copper grades in a bench: three horizontal sections; **b** four vertical sections; **c** blast profile resulting from simulated blast applied to simulated grades; **d** predicted composition of blast profile from simulated blast applied to in-situ grades estimated from samples taken from blast-holes on 8 m spacing

Fig. 18.16 (Left) selected ore volumes based on estimates (Right) actual ore volumes

The first aspect of predicting recovery is the in-situ heterogeneity of the ore and the extent to which it forms contiguous 'parcels' of a size relative to the selection size (capacity and size of loading equipment). The second aspect is the heterogeneity of the ore after it has been subjected to blasting (i.e., the in-situ geological spatial variability and the post-transfer in-situ blast-pile spatial variability).

Figure 18.15 shows horizontal and vertical cross-sections through a simulated bench of dimensions 80 m × 40 m × 12 m (height) simulated copper grades on horizontal planes at the top and bottom of a 12 m bench height and a 6 m mid-plane. The vertical cross-sections of the bench are extremities (0 and 80 m) and intermediate planes at 28 m intervals.

Figure 18.16 shows the assumed contiguous parcels of ore in the blast pile based on estimated in-situ grade values together with the actual (simulated) parcels of ore. A comparison of the two sets of ore volumes in Fig. 18.16 would quantify ore loss and ore dilution. Blast movement sensors, inserted in drill holes and detected in the blast-pile, are widely used to identify post-blast ore parcels. In such cases, this process would quantify the uncertainty associated with the initial placement of sensors based on estimated in-situ ore locations and a grade continuity model.

Among other examples, Goodfellow and Dimitrakopoulos (2017) describe an approach that integrates sources of uncertainties arising from the combined production of several mines. The in-situ orebody uncertainties are integrated with process uncertainties from extraction to processing to marketing as the basis of modelling and stochastically optimising the value chain of a mining complex.

18.7   Conclusion

There is a growing requirement for integrated frameworks for uncertainty quantification in all geologically based applications. Quantified uncertainty and geostatistical methods are increasingly being referenced explicitly in mineral resource and reserve codes. This does not require rewriting the reporting codes but it does mean that there is a need to establish a general accepted framework for the quantification of all sources of uncertainty.

Quantified risk assessments for environmental applications are now required in many jurisdictions for applications such as waste burial and the treatment, storage and disposal of radioactive material. These assessments are required to cover time periods that range from around 200 years for household wastes to thousands of years for the underground storage or disposal of radioactive wastes.

The management of groundwater resources, especially karst systems in environmentally vulnerable coastal areas, requires the integration of flow, extraction, seawater intrusion, contamination from agriculture and other activities.

In these and all such applications the identification and quantification of all sources of uncertainty is critical to ensuring reliable estimation, planning, design and, for resource extraction, production and to managing associated risks. As summarised here, many methods and approaches have been developed by many authors but most are limited to aleatory uncertainty.

The work summarised here provides examples of methods that have been successfully applied to identify and quantify all sources of uncertainty in mineral resource and environmental applications. They provide a contribution to the need, and the increasing requirement, to develop integrated frameworks for uncertainty quantification in all geologically based applications.

Acknowledgements I am grateful to my co-authors of our cited joint publications and particularly to Eulogio Pardo-Igúzquiza with whom I have collaborated for over 20 years.

References

Bandemer H, Gebhart A (2000) Bayesian fuzzy kriging. Fuzzy Sets Syst 112:405–418

Bardossy A, Bogardi I, Kelly WE (1988) Imprecise (fuzzy) information in geostatistics. Math Geol 20:287–311

Bardossy A, Bogardi I, Kelly WE (1990a) Kriging with imprecise (fuzzy) variograms. I: theory. Math Geol 22:63–79

Bardossy A, Bogardi I, Kelly WE (1990b) Kriging with imprecise (fuzzy) variograms. II: application. Math Geol 22:81–94

Bardossy G, Fodor J (2004) Evaluation of uncertainties and risks in geology: new mathematical approaches for their handling. Springer. ISBN: 978-3-642-05833-2

Bedford T, Cooke R (2001) Probabilistic risk analysis: foundations and methods. Cambridge University Press. ISBN: 978-052-1773-20-1

Caers J (2011) Modelling uncertainty in the earth sciences. Wiley-Blackwell. ISBN: 978-111-9992-63-9

Dempster AP (1968) A generalisation of Bayesian inference. J Roy Stat Soc B 30:205–247

Diamond P (1989) Fuzzy kriging. Fuzzy Sets Syst 33:315–332

Dietrich CR, Osborne MR (1991) Estimation of covariance parameters in kriging via restricted maximum likelihood. Math Geol 23(7):655–672

Dimitrakopoulos R, Farrelly CT, Godoy M (2002) Moving forward from traditional optimization: grade uncertainty and risk effects in open-pit design. Trans Inst Min Metall Sect A Min Technol 111:82–88

Dowd PA (1986) Geometrical and geological controls in geostatistical estimation and orebody modelling. In: Ramani RV (ed) Proceedings of 19th APCOM conference, Jostens Publications, pp 81–94. ISSN: 0741-0603; ISBN: 0-87335-058-8

Dowd PA (1994) Geological controls in the geostatistical simulation of hydrocarbon reservoirs. Arab J Sci Eng 19(2B):237–247

Dowd PA (1995) Björkdal gold-mining project, northern Sweden. Trans Inst Min Metall Sect A Min Ind 104:149–163

Dowd PA (1997) Risk in minerals industry projects: analysis, perception and management. Trans Inst Min Metall Sect A Min Ind 106:9–18

Dowd PA, Dare-Bryan PC (2004) Planning, designing and optimising production using geostatistical simulation. In: Proceedings of the international symposium on orebody modelling and strategic mine planning, AusIMM (Melbourne). ISBN: 1-920806-22-9; 321-338

Dowd PA, Scott IR (1984) The application of geostatistics to mine planning in a structurally complex silver/lead/zinc orebody. In: Jones MJ (ed) Proceedings of the 18th APCOM conference; pub. institution of mining and metallurgy, London. ISBN: 0-900488-73-5. 255-264

Dowd PA, Johnstone SAW, Bower J (1989) The application of structurally controlled geostatistics to the Hilton orebodies, Mt. Isa, Australia. In: Weiss A (ed) Proceedings of 21st APCOM conference, society of mining engineers, Colorado, USA. pp 275-285. ISBN 0-87335-079-0

Dowd PA, Pardo-Igúzquiza E (2002) Incorporation of model uncertainty in geostatistical simulation. Geogr Environ Model 6(2):149–171

Dubois D, Prade H (2001) Possibility theory, probability theory and multiple-valued logics: a clarification. Ann Math Artif Intell 32:35–66

Goodfellow R, Dimitrakopoulos R (2013) Algorithmic integration of geological uncertainty in pushback designs for complex multi-process open pit mines. Trans Inst Min Metall Sect A Min Technol 122(2):67–77

Goodfellow R, Dimitrakopoulos R (2017) Simultaneous stochastic optimization of mining complexes and mineral value chains. Math Geosci 49:341–360

Gotway CA (1994) The use of conditional simulation in nuclear waste site performance assessment. Technometrics 36(2):129–141

Helton JC, Johnson JD, Oberkampf WL (2004) An exploration of alternative approaches to the representation of uncertainty in model predictions. Reliab Eng Syst Saf 85:39–71

Hermans T, Caers J, Nguyen F (2014) Assessing the probability of training image-based geological scenarios using geophysical data. In: Pardo-Igúzquiza E, Guardiola-Albert C, Heredia J, Moreno L, Durán J, Vargas-Guzmán J (eds) Mathematics of planet earth. Lecture notes in earth system sciences. Springer, pp 679–682

Hoppe RW (1978) Stekenjokk: a mixed bag of tough geology and good mining and milling practices. Engineering and mining journal operating handbook of underground mining, pp 270–274. ISBN: 0-0709-9928-7

Hora SC (1996) Aleatory and epistemic uncertainty in probability elicitation an example from hazardous waste management. Reliab Eng Syst Saf 54:217–223

JORC Code (2012) Australasian code for reporting of exploration results, mineral resources and ore reserves. http://www.jorc.org

Journel AG (1994) Modelling uncertainty; some conceptual thoughts. In: Dimitrakopoulos R (ed) Geostatistics for the next century; Kluwer quantitative geology and geostatistics series, vol 6, pp 30–43. ISBN: 0-7923-2650-4

Kitanidis PK (1986) Parameter uncertainty in estimation of spatial functions: Bayesian analysis. Water Resour Res 22(4):499–507

Kitanidis PK, Lane RW (1985) Maximum likelihood parameter estimation of hydrologic spatial processes by the Gauss-Newton method. J Hydrol 79(1–2):53–71

Loquin K, Dubois D (2010a) Kriging and epistemic uncertainty. In: Jeansoulin R, Papini O, Prade H, Shockaert S (eds) Methods for handling imperfect spatial information. Studies in fuzziness and soft computing, vol 256. Springer, pp 269-305. ISSN: 1434-9922; ISBN: 978-3-642-14754-8

Loquin K, Dubois D (2010b) Kriging with ill-known variogram and data. In: Deshpande A and Hunter A (eds) Scalable uncertainty management. Lecture notes in computer science, vol 6379. Springer, Berlin, pp 219-235. ISSN: 0302-9743; ISBN: 978-3-642-15950-3

Mardia KV, Marshall RJ (1984) Maximum likelihood estimation of models for residual covariance in spatial regression. Biometrika 71(1):135–146

Matheron G (1975) Hasard, échelle et structure. Ann des Min

Matheron G (1976) Le choix des modèles en géostatistique. In: Guarascio M, David M, Huijbregts C (eds) Advanced geostatistics in the mining industry, NATO A.S.I. Series C: Mathematical and physical sciences, vol 24. D. Reidel Pub. Co, pp 11–27. Print ISBN: 978-940-1014-72-4. Online: 978-940-1014-70-0

Matheron G (1978) Estimer et choisir. Centre de Géostatistique et de Morphologie Mathématique, Fontainebleau. English translation: Hasofer AM (1989) Estimating and Choosing: an essay on probability in practice. Springer. ISBN: 978-3-540-50087-2. Republished 2013 by Presses des Mines, France; ISBN: 978-2-35671-056-7

Oberkampf WL, DeLand SM, Rutherford BM, Diegert KV, Alvin KF (2002) Error and uncertainty in modelling and simulation. Reliab Eng Syst Saf 75:333–357

Oberkampf WL, Helton JC, Joslyn CA, Wojtkiewicz SF, Ferson S (2004) Challenge problems: uncertainty in system response given uncertain parameters. Reliab Eng Syst Saf 85:11–19

Olea RA, Pardo-Igúzquiza E, Dowd PA (2015) Robust and resistant semi-variogram modelling using a generalized bootstrap. J South Afr Inst Min Metall 115:37–44

Omre H (1987) Bayesian kriging—merging observations and qualified guesses in kriging. Math Geol 19(1):25–39

Pardo-Igúzquiza E, Dowd PA (1997a) Statistical inference of covariance parameters by approximate maximum likelihood estimation. Comput Geosci 23(7):793–805

Pardo-Igúzquiza E, Dowd PA (1997b) A case study of model selection and parameter inference by maximum likelihood with application to uncertainty analysis. Non-renew Resour 7(1):63–73

Pardo-Igúzquiza E, Dowd PA (1997c) Maximum likelihood inference of spatial covariance parameters of soil properties. Soil Sci 163(3):212–219

Pardo-Igúzquiza E, Dowd PA (2003) Assessing the uncertainty of spatial covariance parameters of soil properties estimated by maximum likelihood. Soil Sci 168(11):769–782

Pardo-Igúzquiza E, Dowd PA (2013) Comparison of inference methods for estimating semi-variogram model parameters and their uncertainty: the case of small data sets. Comput Geosci 50:154–164

Pardo-Igúzquiza E, Dowd PA, Baltuille JM, Chica-Olmo M (2013) Geostatistical modelling of a coal seam for resource risk assessment. Intern J Coal Geol 112:134–140

Park H, Scheidt C, Fenwick D, Boucher A, Caers J (2013) History matching and uncertainty quantification of facies models with multiple geological interpretations. Comput Geosci 17 (4):609–621

Sankararaman S, Mahadevan S (2011) Model validation under epistemic uncertainty. Reliab Eng Syst Saf 96:1232–1241

Shafer G (1976) A mathematical theory of evidence. Princeton University Press. ISBN: 978-069-1100-42-5

Srivastava RM (1994) Comments on modelling uncertainty: some conceptual thoughts. In: Dimitrakopoulos (ed) Geostatistics for the next century; Kluwer quantitative geology and geostatistics series, vol 6. ISBN: 0-7923-2650-4. 44-45

Srivastava RM (2005) Probabilistic modelling of ore lens geometry: an alternative to deterministic wireframes. Math Geol 37(5):513–544

Verly G, Brisebois K, Hart W (2008) Simulation of geological uncertainty, resolution porphyry copper deposit. In: Proceedings of the eighth geostatistics congress, vol 1, Santiago, Chile. pub. Gecamin Ltd, pp 31–40. ISBN: 978-956-8504-18-2

Winkler RL (1981) Combining probability distributions from dependent information sources. Manag Sci 27(4):479–488

Winkler RL (1996) Uncertainty in probabilistic risk assessment. Reliab Eng Syst Saf 54:127–132

Xu C, Dowd PA (2014) Stochastic fracture propagation modelling for enhanced geothermal systems. Math Geosci 46(6):665–690

Zadeh LA (1965) Fuzzy sets. Inf Control 8:338–353

Zadeh LA (1978) Fuzzy sets as a basis for a theory of possibility. Fuzzy Sets Syst 1:3–28

Zimmerman DL (1989) Computationally efficient restricted maximum likelihood estimation of generalised covariance functions. Math Geol 21(7):655–672

19

Fluctuations in Uncertainty and its Prediction: Distance Methods, Kriging and Stochastic Simulation

Ricardo A. Olea

Abstract A comparative analysis of distance methods, kriging and stochastic simulation is conducted for evaluating their capabilities for predicting fluctuations in uncertainty due to changes in spatially correlated samples. It is concluded that distance methods lack the most basic capabilities to assess reliability despite their wide acceptance. In contrast, kriging and stochastic simulation offer significant improvements by considering probabilistic formulations that provide a basis on which uncertainty can be estimated in a way consistent with practices widely accepted in risk analysis. Additionally, using real thickness data of a coal bed, it is confirmed once more that stochastic simulation outperforms kriging.

19.1 Introduction

In any form of sampling, there is always significant interest in establishing the reliability that may be placed on any conclusions extracted from a sample of certain size. In the earth sciences and engineering, such conclusions can be the extension of a contamination plume or the in situ resources of a mineral commodity. Increases in sample size result in monotonic improvements with diminishing returns: up to measuring the entire population, the benefits increase with the number of observations. In the classical statistics of independent random variables, the number of observations is all that counts. In spatial statistics, however, the locations of the data are also important.

Early on in spatial sampling, it was recognized that sampling distance was a factor in determining the reliability of estimations. However, insurmountable difficulties of incorporating other factors led to the reliability of spatial samplings

R. A. Olea (✉)
U.S. Geological Survey, 12201 Sunrise Valley Drive, Mail Stop 956,
Reston, VA 20192, USA
e-mail: rolea@usgs.gov

being determined solely by geographical distance, particularly for the public disclosure of mineral resources (e.g., USBM and USGS 1976).

Significant advances in the determination of spatial uncertainty did not take
place until the advent of digital computers and the formulation of geostatistics (e.g.,
Matheron 1965). Geostatistics introduced the concept of kriging variance, which
was a significant improvement over the relatively simplistic distance criteria for
determining reliability. The third generation of methods to determine reliability of
spatial sampling came with the development of spatial stochastic simulation shortly
after the formulation of kriging (Journel 1974).

Although there are several reports in the literature about applications of distance
methods (e.g., USGS 1980; Wood et al. 1983; Rendu 2006) and kriging (e.g., Olea
1984; Bhat et al. 2015), the mere fact that distance methods are still being used
indicates that the merits of the geostatistical methods remain unappreciated. This
chapter is an application of the three families of methods for conducting sensitivity
analyses on the reliability of the assessment of geologic resources due to variations
in sample spacing. The simulation formulation given here is novel as it is an
illustrative example used for comparing all three approaches.

19.2 Data

The data in Fig. 19.1 and Table 19.1 of the Appendix will be used to anchor the
presentation. They are thickness measurements for the Anderson coal bed in a
central part of the Gillette coal field of Wyoming taken from a more extensive study
(Olea and Luppens 2014). A conversion factor could have been used to transform

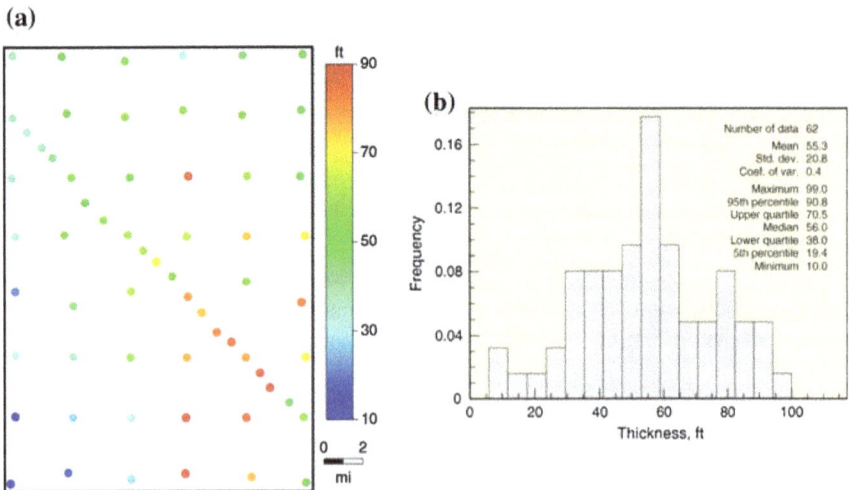

Fig. 19.1 Measurements of thickness for the Anderson coal bed in a central part of the Gillette
coal field, Wyoming, USA: **a** posting of values; **b** histogram

all the thickness values to tonnage, but it was decided to perform the analysis in terms of the attribute actually measured. The reader may want to know, however, that a density of 1,770 short tons per acre-foot for subbituminous coal is a good average value to estimate tonnage values and that the cell size used here is 400 ft by 400 ft.

With resources of more than 200 billion short tons of coal in place, the Gillette coal field is one of the largest coal deposits in the United States (Luppens et al. 2008). There are eleven beds of importance in the field. The Anderson coal bed, in the Paleocene Tongue River Member of the Fort Union Formation, is the thickest and most laterally continuous of the six most economically significant beds. This low sulfur, subbituminous coal has a field average thickness of 45 ft. Hence, it is the main mining target.

19.3 Traditional Uncertainty Assessment

For a long time, the prevailing practice has been the determination of uncertainty in mining assessments based on distance between drill holes. Figure 19.2 shows an example following U.S. Geological Survey Circular 891 (Wood et al. 1983), hereafter referred to as Circular 891. This example uses the drill holes in Fig. 19.1a after eliminating the holes along the diagonal. Circular 891 classifies resources into four categories according to the distance from the estimation location to the closest drill hole:

- 0 to ¼ mi: measured
- ¼ to ¾ mi: indicated
- ¾ to 3 mi: inferred
- More than 3 mi: hypothetical

Fig. 19.2 Classification of in situ resources according to Circular 891 for the data in Fig. 19.1a after eliminating the drill holes along the diagonal

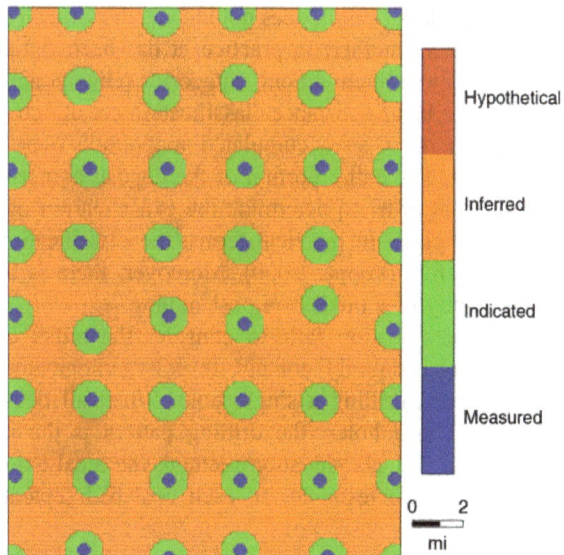

Classification schemes like this are fairly simple and gained popularity prior to the advent of computers. Evaluating the degree of uncertainty of a magnitude or an event is the domain of statistics (e.g., Caers 2011). The standard approach for analyzing uncertainty consists of listing all possible values or events and then assigning a relative frequency of occurrence. A simple example is the tossing of a coin, where the outcomes are head and tail. For a fair coin, these two events occur with the same frequency, which is called probability when normalized to vary from 0 to 1. The same concept can be applied to any event or attribute, including coal bed thickness. For example, the outcome at a site not yet drilled could be modeled as the following random variable:

- 5–10 ft, probability 0.3
- 10–15 ft, probability 0.4
- 15–21 ft, probability 0.2
- 21–28 ft, probability 0.1

Note that the sum of the probabilities of all possible outcomes is 1.0. Random variables rigorously allow answering multiple questions about unknown magnitudes, in this case, the likely thickness to penetrate. A sample of just three assertions would be: (a) coal will certainly be intersected because the value zero is not listed among the possibilities; (b) it is more likely that the intersected thickness will be less than 15 ft than greater than 15 ft; and (c) odds are 6 to 4 that the thickness will be between 10 and 21 ft, or to put it differently, the 11 ft interval between 10 and 21 ft has a probability of 0.6 of containing the true thickness. These are the standard concepts and tools used universally in statistics to characterize uncertainty.

The classification system established by Circular 891 does not use probabilities and lacks the predictive power of a random variable approach. In particular,

- The classification uses an ordinal scale (e.g., Urdan 2017), supposedly ranked, but the classification does not indicate how much more uncertain one category is relative to another. In practice, it has been found that errors may not be significantly different among categories (Olea et al. 2011).
- The results of a distance classification are difficult to validate. The tonnage in a class denotes an accumulated magnitude over an extensive volume of the deposit. The entire portion of the deposit comprising a class would have to be mined in order to determine the exact margin of error in the classification for such a class. In practical terms, the classification is not falsifiable, thus it is unscientific (Popper 2000). Moreover, there is little value in determining the reliability of a prediction post mining.
- The classification fails to consider the effect of geologic complexity. Coal deposits ordinarily contain several geologically different beds that may be penetrated drilling a single hole. When all beds are penetrated by the same vertical drill holes, the drilling pattern is the same for all beds. Using the Circular 891 classification method, the areal extension of each category is the same for the resources of each coal bed separately and for the accumulated

resources considering all coal beds, while logic indicates that the extension of true reliability classes should be all different.

- For similar reasons, in a multi-seam deposit, increasing the drilling density results in the same reduction in uncertainty for all coalbeds, which is also unrealistic.
- The number of methods for estimating resources is continuously growing, hopefully for the better. Considering that not all methods are equally powerful, independently of the data, different methods offer varying degrees of reliability. The uncertainty denoted by the Circular 891 classification is insensitive to the methods used in the calculation of the tonnage. For example, inferred resources remain as inferred resources independently of the nature and quality of the methods used in the assessment.

Despite these drawbacks and the formulation of the superior alternatives below, Circular 891 and similar approaches remain the prevailing methods worldwide for the public disclosure of uncertainty in the assessment of mineral resources and reserves (JORC 2012; CRIRSCO 2013).

19.4 Kriging

Kriging is a family of spatial statistics methods formulated for the improvement in the reporting of uncertainty and in the estimation of the attributes of interest themselves. Although it is possible to establish links between kriging and other older estimation methods in various disciplines, mining was the driving force behind the initial developments of kriging and other related methods collectively known today as geostatistics (Cressie 1990).

Kriging is basically a generalization of minimum mean square error estimation taking into account spatial correlation. Kriging provides two numbers per location (s_o) conditioned to some sample of the attribute $(z(s_i), i = 1, 2, \ldots, N)$: an estimate of the unknown value $(z^*(s_o))$ and a standard error $(\sigma(s_o))$. The exact expression for these results depends on the form of kriging. For ordinary kriging, the most commonly applied form and the one used here, the equations are:

$$z^*(s_o) = \sum_{i=1}^{n} \lambda_i \cdot z(s_i) \tag{19.1}$$

$$\sigma^2(s_o) = \left(\sum_{i=1}^{n} \lambda_i \cdot \gamma(s_o, s_i) \right) - \mu \tag{19.2}$$

where:

$n \leq N$ is a subset of the sample consisting of the observations closest to s_o;
$\gamma(\mathbf{d})$ is the semivariogram, a function of the distance $\mathbf{d}$ between two locations;

λ_i is a weight determined by solving a system of linear equations comprising semivariogram terms; and

μ is a Lagrange multiplier, also determined by solving the same system of equations.

The method presumes knowledge of the function characterizing the spatial correlation between any two points, which is never the case. A structural analysis must be conducted before running kriging to estimate this function: a covariance or semivariogram. The semivariogram can be regarded as a scaled distance function. The weights and the Lagrange multiplier depend on the semivariogram for multiple drill-hole to drill-hole distances and estimation location to drill-hole distances. For details, see for example Olea (1999).

The two terms, $z^*(\mathbf{s}_o)$ and $\sigma^2(\mathbf{s}_o)$, are the mean and the variance of the random variable modeling the uncertainty of the true value of the attribute $z(\mathbf{s}_o)$, terms that are compatible with all that is known about the attribute through the sample of size N. Variance is a measure of dispersion, in this case, dispersion of possible values around the estimate, which is the most likely value. Hence, changing the sample, a sensitivity analysis of kriging variance is a sensitivity analysis of variations in uncertainty due to changes in the sampling scheme. From Eq. 19.2, the kriging variance does not depend directly on the observations. The dependence is only indirect through the semivariogram, which is based on the data. Considering that there is one true semivariogram per attribute, changes in adequate sampling should not result in significant changes in the estimated semivariogram, which is kept constant. This independence between data and standard error facilitates the application of kriging to the sensitivity analysis in the reliability of an assessment due to changes in sampling strategy because mathematically actual measurements are not necessary to calculate standard errors; the modeler only has to specify the semivariogram and the sampling locations.

Figure 19.3 shows the set of estimated semivariogram values obtained using the sample in Fig. 19.1 plus a model fitting the points for the purpose of having valid semivariogram values for any distance. In this case, the fitted curve is called a spherical model with a nugget of 20 sq ft, sill of 595 sq ft and a spatial correlation range of 88,920 ft. Geologically, the nugget is related to the variance of short scale fluctuations; the sill is of the same order of magnitude as the sample variance, and the correlation range is equal to half the average geographical size of the anomalies. For details on structural analysis, see for example Olea (2006).

Figure 19.4 shows the results of applying ordinary kriging to the sample in Fig. 19.1a and Table 19.1 in the Appendix. As expected, the standard error is zero at the drill holes because there is no uncertainty where measurements have been taken.

Although kriging can analyze any configuration, Fig. 19.5 only relates to additions or eliminations to the basic sample in Fig. 19.1a. Values along the diagonal were used only for modeling the semivariogram and producing Fig. 19.4. Figure 19.5a also has every other row and column eliminated. Estimates could be produced for the first two configurations because thickness is known at each drill hole. The other maps were produced by interpolating locations in the sample with

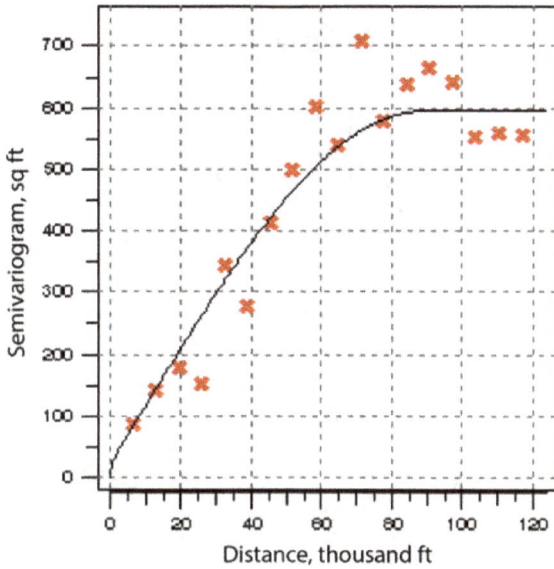

Fig. 19.3 Semivariogram for the Anderson coal bed thickness. The crosses denote estimated values and the curve is a model fitting the values

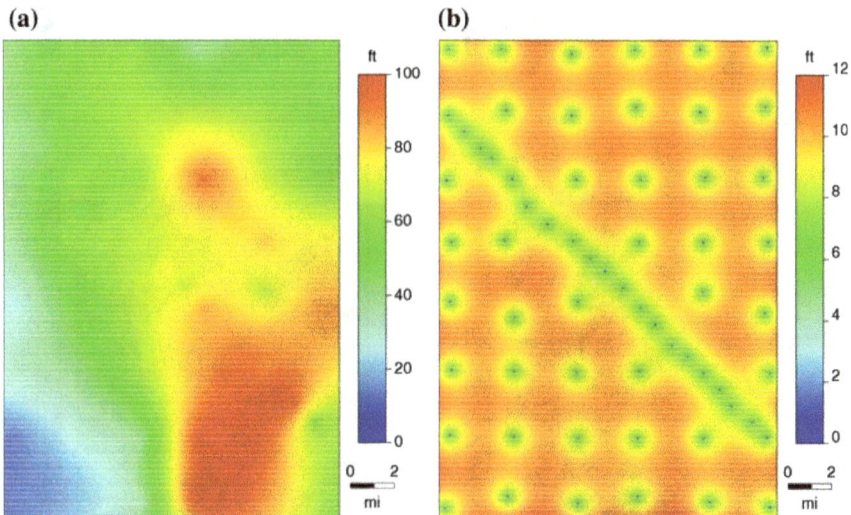

Fig. 19.4 Ordinary kriging maps for the Anderson coal bed in a central part of the Gillette coal field (Wyoming) using the sample in Fig. 19.1: **a** thickness; **b** standard error

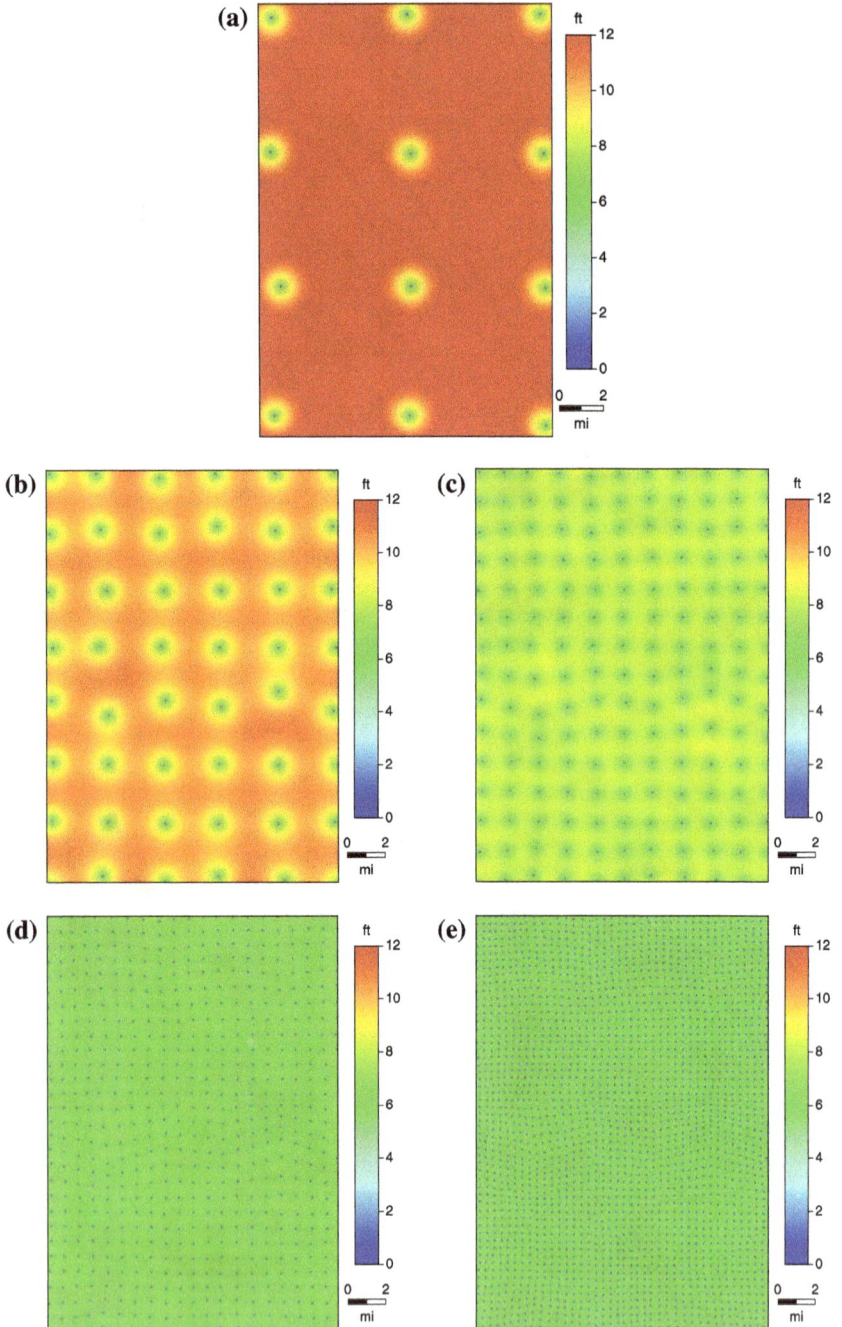

Fig. 19.5 Ordinary kriging standard error for the same configuration in Fig. 19.2 for several average spacings: **a** 6 mi; **b** 3 mi; **c** 1.5 mi; **d** 3/4 mi; **e** 3/8 mi

the next largest spacing; it is only possible to produce the standard error map for Fig. 19.5c–e.

The similarity between Figs. 19.2 and 19.5b may lead to incorrect conclusions. Although the location and extension of similar colors are approximately the same, what is important is the meaning of the colors. Figure 19.2 does not provide any numerical information that can be associated with the accuracy and the precision of the estimated values. In Fig. 19.5b the numbers are standard errors, a direct measurement of estimation reliability. In other more irregular configurations, there will not be similarity in color patterns no matter how the colors are selected. For example, by expanding the boundary of the study area, Fig. 19.6 shows how the Circular 891 classification is totally insensitive to the fact that, along the periphery, there is an increase in uncertainty because the data are now to one side, not surrounding the estimation locations. Instead, kriging accounts for the fact that extrapolation is always a more uncertain operation than interpolation, an important capability when accounting for boundary effects.

Kriging is able to provide random variables for the statistical characterization of uncertainty if the modeler is willing to introduce a distributional assumption. $z^*(\mathbf{s}_o)$ and $\sigma^2(\mathbf{s}_o)$ are the mean and the variance of the distribution of the random variable providing the likely values for $z(\mathbf{s}_o)$. These parameters are necessary but not sufficient to fully characterize any distribution. However, this indetermination can be eliminated by assuming a distribution that is fully determined with these two parameters. Ordinarily, the distribution of choice is the normal distribution, followed by the lognormal. The form of the distribution does not change by subtracting $z(\mathbf{s}_o)$ from all estimates. As the difference $z^*(\mathbf{s}_o) - z(\mathbf{s}_o)$ is the estimation error, the distributional assumption also allows characterizing the distribution for the error at $\mathbf{s}_o$.

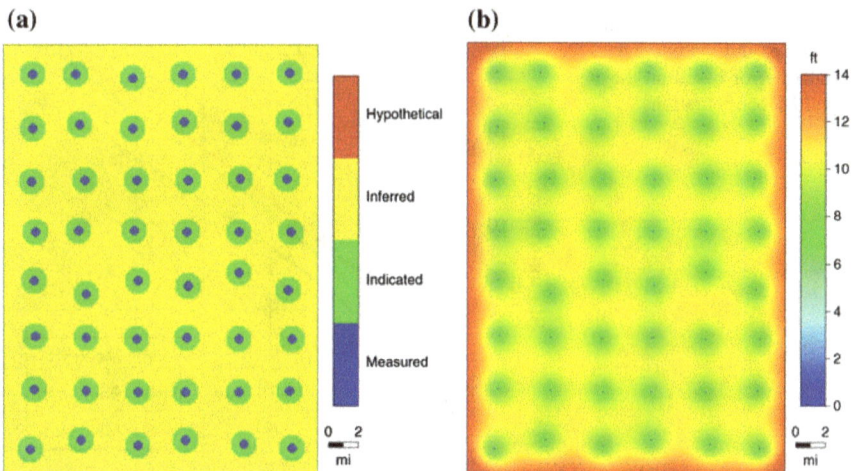

Fig. 19.6 Comparison of results when expanding the boundaries of the study area: a Circular 891 classification; b ordinary kriging standard error

Kriging with a distribution for the errors overcomes all the disadvantages of the distance methods listed in the previous section:

- It is possible to calculate the probability that the true value of the attribute lies in any number of intervals. Probabilities are a form of a ratio variable, for which zero denotes an impossible event and, say, a 0.2 probability denotes twice the likelihood of occurrence of an event than 0.1.
- Validation is modular. An adequate theory assures that, on average, $z^*(\mathbf{s}_o)$ and $\sigma^2(\mathbf{s}_o)$ are good estimates of reality. Yet, as illustrated by an example in the last Section, if going ahead with validation of the uncertainty modeling primarily to check the adequacy of the normality assumption, it is not necessary to validate all possible locations throughout the entire deposit to evaluate the quality of the modeling.
- The effect of complexity in the geology is taken into account by the semivariogram.
- In general, the thickness of every coal bed or the accumulated values of thickness for several coal beds has a different semivariogram. Thus, even if the sampling configuration is the same, the standard error maps will be different.
- The characterization of uncertainty is specific to the estimation method because the results are valid only for estimated values using the same form of kriging used to generate the standard errors.

Figure 19.7 summarizes the results of the maps in Fig. 19.5. Display of the 95th percentile is based on the assumption that all random variables follow normal distributions. The curves clearly outline the consequences of varying the spacing in

Fig. 19.7 Sensitivity of ordinary kriging to spacing of mean standard error, its 95th percentile, and the maximum standard error based on the Fig. 19.5 configurations

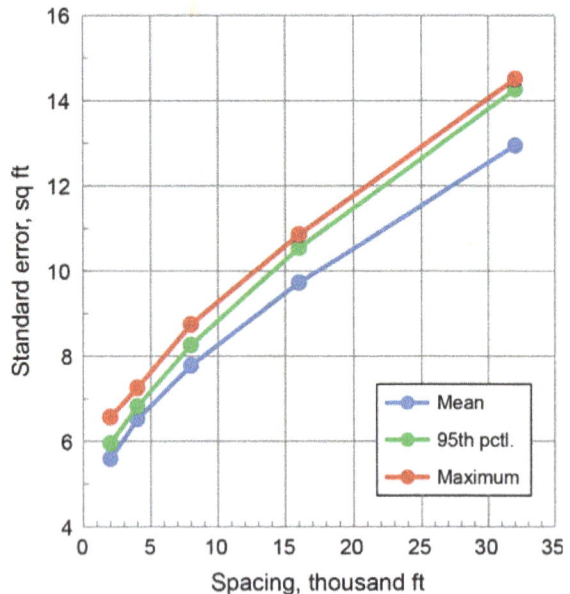

a square sampling pattern from 2,000 to 32,000 ft. So, for example, if it is required that all estimates in the study area must have a standard error less than 10 ft, then the maximum spacing must be at most 12,500 ft. The validity of the results, however, is specific to the attribute and sampling pattern: thickness of the Anderson coal bed investigated with a square grid. Any change in these specifications requires preparation of another set of curves.

19.5 Stochastic Simulation

Despite limited acceptance, the kriging variance has been in use for a while in the sensitivity analysis of uncertainty to changes in sampling distances and configurations (e.g., Olea 1984; Cressie et al. 1990). Kriging, like any mathematical method, has been open to improvements. One result has been the formulation of another family of methods: stochastic simulation.

Relative to the topic of this chapter, stochastic simulation offers two improvements: (a) it is no longer necessary to assume the form for the distribution providing all possible values for the true value of the attribute $z(s_o)$; and (b) the standard error is sensitive to the data.

As seen in Fig. 19.4, for every attribute and sample, kriging produces two maps, a map of the estimate and a map of the standard error. The idea of stochastic simulation is to characterize uncertainty by producing instead multiple attribute maps, all compatible with the data at hand and each representing one possible outcome of reality—realization, for short. From among the many available methods of geostatistical simulation, sequential Gaussian simulation has been chosen for this study because of its simplicity, versatility and efficiency (Pyrcz and Deutsch 2014). Figure 19.8 shows four simulated realizations, each of which is a possible reality in the sense that the values have the same statistics and spatial statistics (semivariogram) and the simulation reproduces the known sample values (i.e., the sample used to prepare Fig. 19.5b).

Generation of significant results needs preparation of more realizations than the four in Fig. 19.8. An estimation of uncertainty requires summarizing the fluctuations from realization to realization, either at local or global scales. Figure 19.9 is an example of local fluctuation summarizing all values of thickness at the same location for 100 realizations. This histogram is the numerical characterization of uncertainty through a random variable. There is one random variable for each of the 57,528 pixels (cells) comprising each realization. As clearly implied by the selected values in the tabulation, this collection of 100 maps provides multiple predictions of the true thickness value that should be expected at this location. For example, the most likely value (mean) is 65.75 ft; the standard error is 13.47 ft; and there is a 0.95 probability that the coal bed will be less than 87.8 ft thick.

Maps can be generated for various statistics across the study area to display fluctuations in their values. Figure 19.10 shows a map of the mean and a map of the standard error. Note that the map for the mean is quite similar to the ordinary kriging map in Fig. 19.4a. More importantly, the maps for the standard errors in

(a)

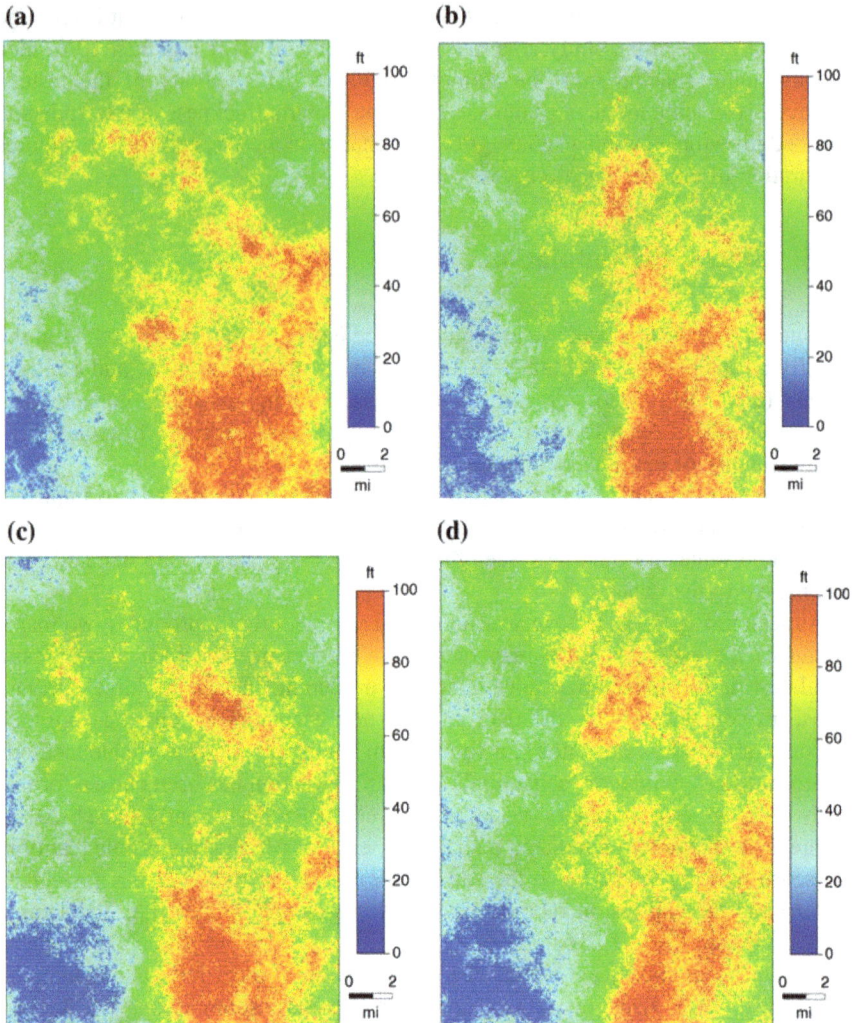

(b)

(c)

(d)

Fig. 19.8 A sample of four sequential Gaussian realizations using the same data used in the preparation of Fig. 19.5b

Figs. 19.5b and 19.10b are significantly different. The differences in the standard errors are primarily the result of the dependency of the standard error not only on the semivariogram and the drill hole locations, but also on the values of thickness as well. For example, comparing Figs. 19.1a and 19.10b, despite the regularity in the drilling, there is less uncertainty in the southwest corner where all values are low as well as in the south central part where all values are consistently high.

Production of a display of the standard error equivalent to that in Fig. 19.5 is more challenging now that the standard deviation must be extracted from multiple realizations and the preparation of each realization requires a value at each drill hole

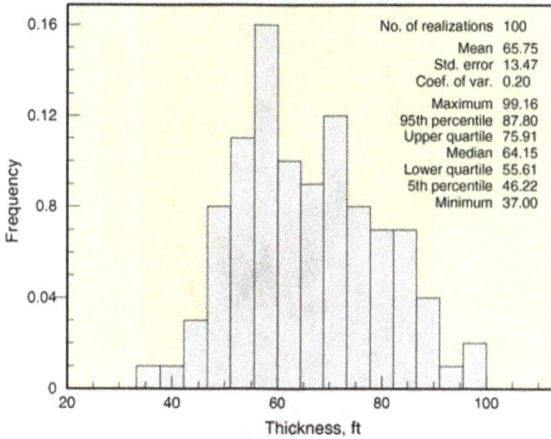

Fig. 19.9 Example of the numerical approximation to the random variable modeling uncertainty in the value of thickness at a site not yet drilled

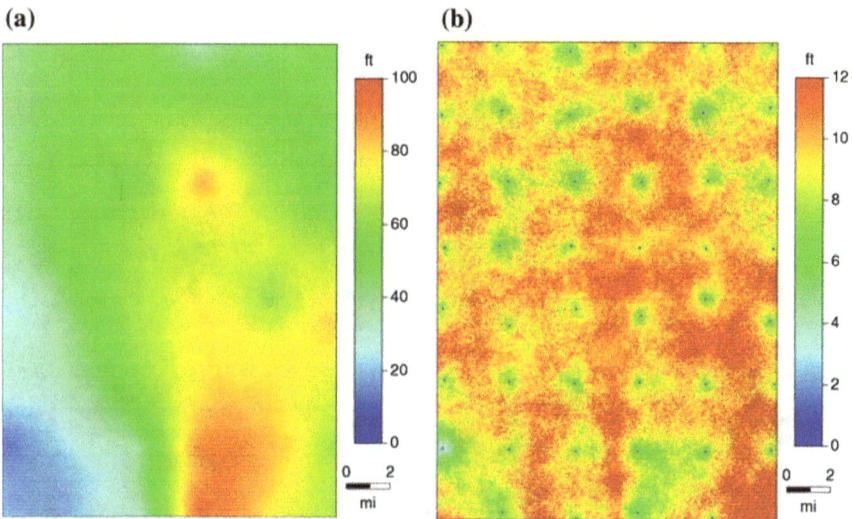

Fig. 19.10 Anderson coal bed thickness according to 100 sequential Gaussian simulations: **a** expected value of thickness; **b** standard error

in the configuration of interest to complete the analysis. Figure 19.11 shows the equivalent results to Fig. 19.5 for the same drill holes, but now produced after applying sequential Gaussian simulation. The additional data necessary to prepare the maps in Fig. 19.11c–e where obtained by randomly selecting 10 of the 100 realizations used to prepare the maps in Figs. 19.8 and 19.10. The data for the hypothetical drill holes were taken from the values at the collocated nodes in these selected 10 realizations, thus obtaining 10 datasets consisting partly of the 48 actual data in Fig. 19.11b plus the artificial data obtained by "drilling" the realizations.

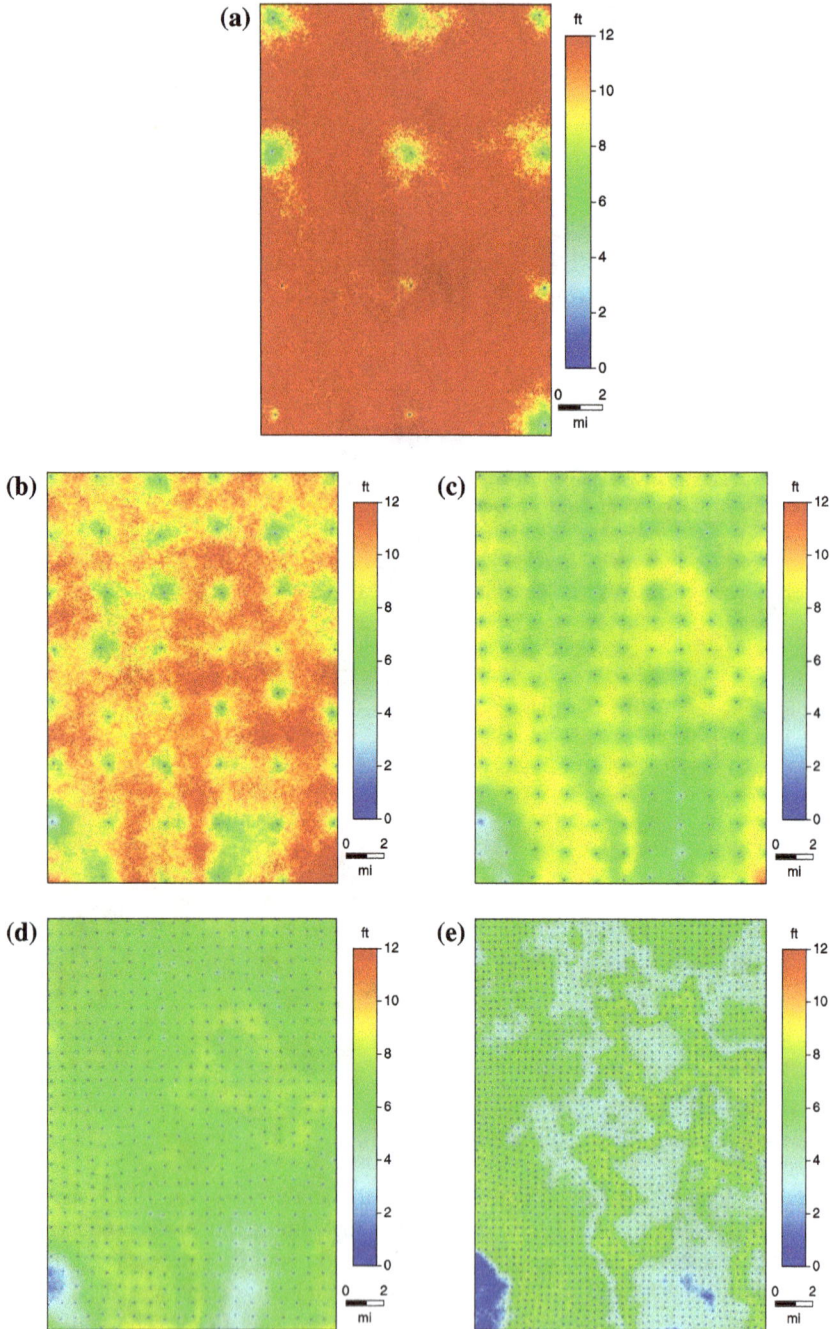

Fig. 19.11 Sequential Gaussian simulation standard error for the same configuration on Fig. 19.2 for several average spacings: **a** 6 mi; **b** 3 mi; **c** 1.5 mi; **d** 3/4 mi; **e** 3/8 mi

Fig. 19.12 Sequential
Gaussian simulation
sensitivity to spacing of mean
standard error, its 95th
percentile, and the maximum
standard error based on the
configurations in Fig. 19.11

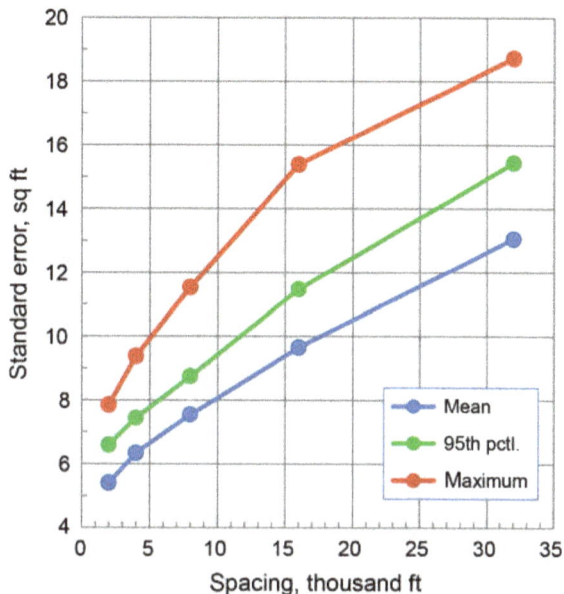

Finally, each dataset was used to generate 100 realizations, for a total of 1,000 realizations per configuration. As mentioned for Fig. 19.10b, despite the regularity of the drill hole pattern, the fluctuations in standard error are no longer completely determined by the drilling pattern.

Figure 19.12 is the summary equivalent to that in Fig. 19.7. Considering the completely different methodologies behind both sets of curves, the results are quite similar, particularly the curves for the mean standard error, which are almost identical. The more extreme standard errors of the sequential Gaussian simulation are larger than those for ordinary kriging in the case of the 95th percentile and the maximum value. The remaining question is: Which approach produces the most realistic forecasts of uncertainty?

19.6 Validation

Figure 19.13 provides an answer to the question above in terms of percentiles. A percentile is a number that separates a set of values into two groups, one below and the other one above the percentile. The percentage of values below gives the name to the percentile. For example, in Fig. 19.9, the value 46.22 ft separates the 100 values of thickness into two classes, those below and those equal to or above 46.22 ft. It turns out that only 5 of the 100 values are below 46.22 ft. Hence, 46.22 ft is the 5th percentile of that dataset. Accepting only integer values of percentages, there are 99 percentiles in any dataset. The quality of a model of uncertainty can be validated by checking the proportion of true values that are actually below the percentiles of the prediction random variables collocated with data not used in the

(a)

(b)

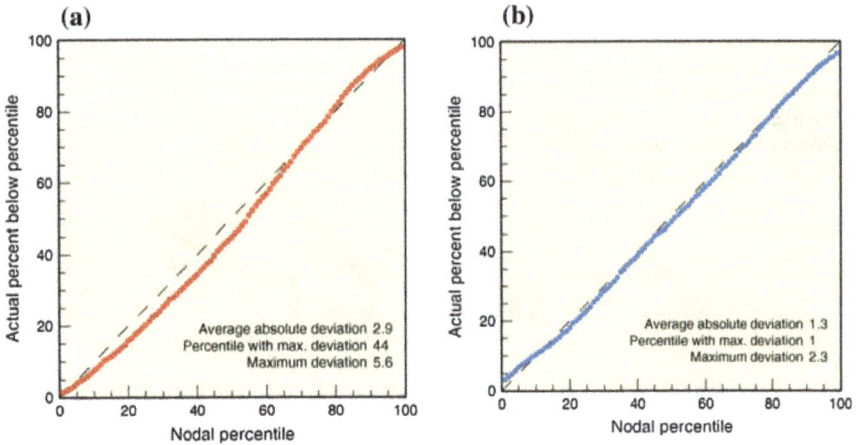

Fig. 19.13 Validation of the uncertainty predictions made for the 3 mi spacing samples: **a** ordinary kriging; **b** sequential Gaussian simulation

modeling. One of the reasons for selecting the Anderson coal thickness for the study is that there are much more data than the 48 values used to generate the realizations, a generous set of 2,136 additional values to be precise. This larger number of values has been used for checking the accuracy of the percentiles, not only the 5th percentile, but all 99 percentiles. In the graphs, the actual percentage shows, on average, the proportion of times the true value was below the percentile of a random variable at the location of a censored measurement. For example, in Fig. 19.13a, 641 times out of 2,136 (i.e., 30%) the true value was indeed below the 35th percentile. Ideally, all dots should lie along the main diagonal. The clear winner is sequential Gaussian simulation.

19.7 Conclusions

Distance methods, kriging and stochastic simulation rank, in that order, in terms of increasing detail and precision of the information that they are able to provide concerning the uncertainty associated to any spatial resource assessment.

The resource classification provided by distance methods is completely independent of the geology of the deposit and the method applied to calculate the mineral resources. The magnitude of the resource per class has no associated quantitative measure of the deviation that could be expected between the calculated resource and the actual amount in place.

The geostatistical methods of kriging and stochastic simulation base the modeling on the concept of random variable used in statistics, which allows the same type of probabilistic forecasting used in other forms of risk assessments. Censored data were used for validating the accuracy of the probabilistic predictions that can be made using the geostatistical methods. The results were entirely satisfactory, particularly in the case of stochastic simulation.

Acknowledgements This contribution completed a required review and approval process by the U.S. Geological Survey (USGS) described in Fundamental Science Practices (http://pubs.usgs. gov/circ/1367/) before final inclusion in this volume. I wish to thank Brian Shaffer and James Luppens (USGS), Peter Dowd (University of Adelaide) and Josep Antoni Martín-Fernández (Visiting Fulbright Scholar, USGS) for suggestions leading to improvements to earlier versions of the manuscript.

Appendix

See Table 19.1.

Table 19.1 Thickness data. ID = identification number; Thick. = thickness; ft = feet

ID	Easting (ft)	Northing (ft)	Thick. (ft)	ID	Easting (ft)	Northing (ft)	Thick. (ft)
2	431,326	1,316,298	49.0	39	399,741	1,236,607	77.0
3	398,753	1,316,124	32.0	40	432,107	1,236,582	70.5
4	352,156	1,316,015	37.0	41	384,280	1,236,527	58.0
5	365,531	1.315,818	49.0	42	415,737	1,236,459	78.0
7	382,816	1,314,601	55.0	44	352,743	1,221,026	10.0
9	430,850	1,301,568	48.0	45	368,483	1,220,742	26.0
10	398,805	1,301,506	57.0	46	431,473	1,220,645	59.0
11	352,234	1,299,533	37.0	47	399,596	1,220,598	92.0
12	366,769	1,300,871	50.0	48	415,871	1,220,477	86.0
13	414,876	1,300,240	56.0	49	384,411	1,220,477	32.0
14	382,892	1,299,775	58.0	51	367,180	1,206,180	17.0
16	416,097	1,284,247	60.0	52	399,353	1,205,960	99.0
17	430,593	1,284,243	47.0	53	417,304	1,204,922	76.0
18	400,291	1,284,132	87.0	54	384,456	1,204,470	28.0
19	384,138	1,283,859	53.0	55	432,027	1,203,507	52.0
20	368,123	1,283,849	56.0	56	351,466	1,203,245	11.0
21	351,956	1,283,728	36.0	123	356,115	1,295,788	35.0
23	366,138	1,268,773	55.0	145	360,095	1,291,759	38.0
24	383,559	1,268,661	60.0	166	362,980	1,289,047	42.0
25	431,915	1,268,363	70.0	216	371,863	1,277,272	50.0
26	415,962	1,268,347	75.0	234	377,019	1,272,660	57.0
27	399,884	1,268,270	63.0	282	387,755	1,264,534	60.0
28	352,933	1,268,254	34.0	299	391,477	1,261,727	70.0
30	352,738	1,253,951	21.0	318	395,814	1,257,798	58.0
31	384,499	1,253,969	62.0	380	403,832	1,248,290	75.0
32	400,076	1,252,554	79.0	406	407,848	1,243,143	81.0
33	415,868	1,256,420	57.0	427	411,790	1,240,470	84.0
34	368,579	1,250,159	44.0	470	419,690	1,232,449	92.0
35	430,979	1,251,072	81.5	497	422,447	1,228,465	90.0
37	352,979	1,237 155	30.0	512	427,604	1,224,598	46.0
38	368,493	1,236,862	37.0	1001	415,000	1,316,000	45.0

References

Bhat S, Motz LH, Pathak C, Kuebler L (2015) Geostatistics-based groundwater-level monitoring network design and its application to the Upper Floridan aquifer, USA. Environ Monit Assess 187(1):4183, 15

Caers J (2011) Modeling uncertainty in the earth sciences. Wiley-Blackwell, Chichester, UK, p 229

Cressie N (1990) The origins of kriging. Math Geol 22(3):239–252

Cressie N, Gotway CA, Grondona MO (1990) Spatial prediction from networks. Chemometr Intell Lab Syst 7(3):251–271

CRIRSCO (Combined Reserves International Reporting Standards Committee) (2013) International reporting template for the public reporting of exploration results, mineral resources and mineral reserves, pp 41. http://www.crirsco.com/templates/international_reporting_template_november_2013.pdf

JORC (Joint Ore Reserves Committee) (2012) Australasian code for reporting of exploration results, mineral resources and ore reserves. http://www.jorc.org/docs/jorc_code2012.pdf

Journel AG (1974) Geostatistics for conditional simulation of ore bodies. Econ Geol 69(5):673–687

Luppens JA, Scott DC, Haacke JE, Osmonson LM, Rohrbacher TJ, Ellis MS (2008) Assessment of coal geology, resources, and reserves in the Gillette coalfield, Powder River Basin, Wyoming. U.S. Geological Survey, Open-File Report 2008-1202, pp 127. http://pubs.usgs.gov/of/2008/1202/

Matheron G (1965) Les variables régionalisées et leur estimation; Une application de la théorie des fonctions aléatories aux Sciences de la Nature. Masson et Cie, Paris, pp 305

Olea RA (1984) Sampling design optimization for spatial functions. J Int Assoc Math Geol 16 (4):369–392

Olea RA (1999) Geostatistics for engineers and earth scientists. Kluwer Academic Publishers, Norwell, MA, p 303

Olea RA (2006) A six-step practical approach to semivariogram modeling. Stoch Env Res Risk Assess 39(5):453–467

Olea RA, Luppens JA (2014) Modeling uncertainty in coal resource assessments, with an application to a central area of the Gillette coal field, Wyoming. U.S. Geological Survey Scientific Investigations Report 2014-5196, pp 46. http://pubs.usgs.gov/sir/2014/5196/

Olea RA, Luppens JA, Tewalt SJ (2011) Methodology for quantifying uncertainty in coal assessments with an application to a Texas lignite deposit. Int J Coal Geol 85(1):78–90

Popper KR (2000) The logic of scientific discovery. Routledge, London, p 480

Pyrcz MJ, Deutsch CV (2014) Geostatistical reservoir modeling, 2nd edn. Oxford University Press, New York, p 433

Rendu JM (2006) Reporting mineral resources and mineral reserves in the United States of America—Technical and regulatory issues. In: Proceedings, Sixth International Mining Geology Conference: Australasian Institute of Mining and Metallurgy, pp 11–20

Urdan TC (2017) Statistics in plain English. Routledge, New York, p 265

USBM and USGS (U.S. Bureau of Mines and U.S. Geological Survey) (1976) Coal resource classification system of the U.S. Bureau of Mines and U.S. Geological Survey. Geological Survey Bulletin 1450-B, pp 7. https://pubs.usgs.gov/bul/1450b/report.pdf

USGS (U.S. Geological Survey) (1980) Principles of resource/reserve classification for minerals. U.S. Geological Survey Circular 831, pp 5. http://pubs.usgs.gov/circ/1980/0831/report.pdf

Wood GH Jr, Kehn TM, Carter MD, Culbertson WC (1983) Coal resources classification system of the U.S. Geological Survey. U.S. Geological Survey Circular 891, pp 65. http://pubs.usgs.gov/circ/1983/0891/report.pdf

20

Estimating Deposit Growth of Molybdenum through a Nonlinear Model

John H. Schuenemeyer, Lawrence J. Drew and James D. Bliss

Abstract In the study of molybdenum deposits and most other minerals deposits, including copper, lead and zinc, there is speculation that most undiscovered ore results from an increase (or "growth") in the estimated size of a known deposit due to factors such as exploitation and advances in mining and exploration technology, rather than in discovering wholly new deposits. The purpose of this study is to construct a nonlinear model to estimate deposit "growth" for known deposits as a function of cutoff grade. The model selected for this data set was a truncated normal cumulative distribution function. Because the cutoff grade is commonly unknown, a model to estimate cutoff grade conditioned upon the deposit grade was constructed using data from 34 deposits with reported data on molybdenum grade, cutoff grade, and tonnage. Finally, an example is presented.

Keywords Porphyry molybdenum · Deposit growth · Cutoff grade
Truncated cumulative distribution model fitting and estimation · Confidence and prediction intervals for nonlinear estimation

20.1 Introduction

Initial estimates of a mineral deposit size based on limited data usually underestimate the ultimate size of a mineral deposit, often by a significant amount. The initial size estimate may be of only marginal interest but the size estimate after some exploration and development can be of significant interest. The steps in this process are the subject of this chapter. "Mineral resources" are defined as concentrations or occurrences of material of economic interest in or on the Earth's crust in such form, quality, and quantity that there are reasonable prospects for eventual economic extraction (Zientek and Hammarstrom 2014), and the term "mineral reserves" is restricted to the economically mineable part of a mineral resource.

J. H. Schuenemeyer (✉) · L. J. Drew · J. D. Bliss
Southwest Statistical Consulting, LLC, Cortez, CO, USA
e-mail: jackswsc@q.com

The reported size of known mineral or oil and gas deposit reserves recorded in the mining literature typically increases through time as subsequent development drilling and mining enlarge the deposit's footprint. This phenomenon is referred to as "deposit growth". In a sense, a deposit is never finished "growing" until it is completely mined out. Research on the growth of a deposit's reserves has been a topic of investigation for many years within the United States Geological Survey. Drew (1997) illustrated the growth of oil and gas fields over time in the United States and determined that a large percentage of the ultimate production of a region could come from deposit growth, if the forecast was made early enough in the discovery process. Long (2008) defined reserve growth as the ratio of current reserves plus past production to original reserves. He examined reserve growth in porphyry copper deposits and found that about 20% of porphyry copper mines in the Western Hemisphere had experienced reserve growth of a factor of 10 or better over initial reserves. Reserve growth at these mines added reserves comparable in size to reserves added through discovery of new deposits during the same time period.

Three variables are required to estimate the ultimate size of a deposit: (1) the grade of the deposit, (2) cutoff grade of the deposit, and (3) associated tonnage of ore at successive points in the development of the deposit (Long 2008). The grade of a deposit is defined as the relative quantity of ore mineral within the orebody, typically expressed as a percentage (or g/t). The grade may vary across an orebody, but commonly an average grade may be applied to the orebody as a whole. A cutoff grade is the lowest grade of mineralized material that qualifies as economically mineable and available in a given deposit (Committee for Mineral Reserves International Reporting Standards 2006). Mined material with a grade below the cutoff grade is not processed into metal but is set aside. As deposit development and mining progress, over time the cutoff grade usually declines in an orderly manner. Tonnage is typically reported in metric tons (mt) and includes the mass of total production, reserves and resources of pre-mined material.

The purpose of this study was to construct a nonlinear model to estimate the incremental deposit "growth" for known mineralized areas as a function of cutoff grade, using porphyry molybdenum deposits as an example. Porphyry molybdenum deposits are related to granitic plutons, mostly of Tertiary age, and are formed by hydrothermal fluids associated with the emplacement of granites. They typically occur as large tonnage, low-grade deposits that are commonly mined using open-pit methods.

Two issues must be addressed to predict porphyry molybdenum deposit growth. The first is that, in many instances, the cutoff grade is not available for a given deposit and thus must be estimated. Thus, the first part of this study uses the known molybdenum grade of a deposit to predict probable cutoff grade. The second part of this study in turn uses this predicted cutoff grade to estimate deposit growth as a function of cutoff grade. Two data sets were used in this study. Nearly all porphyry molybdenum deposits used in this study are for unworked deposits; that is, deposits that have been delineated by drilling but are yet unmined. The first data set (Appendix 1) consists of 34 porphyry molybdenum deposits used to model molybdenum cutoff grade in percent (COG) as a function of molybdenum deposit grade, also expressed in percent. The second data set (Appendix 2) is used to model the deposit growth as a function of cutoff grade. The references to Appendices 1 and 2 are Barnes et al. (2009), Baudry (2009), Becker et al. (2009), British Columbia Ministry of Energy and Mines (2012, 2014a, b), Chen and Wang (2011), Ewert et al. (2008), General Moly (2012), Geological Survey of Finland (2011), Geoscience Australia (2012), Kramer (2006), Lowe et al. (2001), Ludington and Plumlee (2009), Mercator Minerals (2011), Mindat.org (1992, 2011), Nanika Resources Inc (2012), Northern Miner (2010), Raw Minerals Group (2011), RX Exploration Inc (2010), Singer et al. (2008), Smith (2009), Taylor et al. (2012), Thompson Creek Metals Company Inc (2011), TTM Resources Inc (2009), US Geological Survey (2011), Wu et al (2011), Yukon Geological Survey (2005). The authors know of no subset of publications that cite the deposits presented in Appendices 1 and 2.

20.2　Cutoff Grade as a Function of Deposit Grade

The first and most straightforward of the two models to analyze is the relationship between molybdenum cutoff grade (Mo COG, %) as a function of molybdenum deposit grade (Mo Grade, %) for the 34 deposits shown in Appendix 1. A scatter plot between these two variables plus a fitted linear regression line, 95% confidence intervals, and 95% prediction intervals are shown in Fig. 20.1.

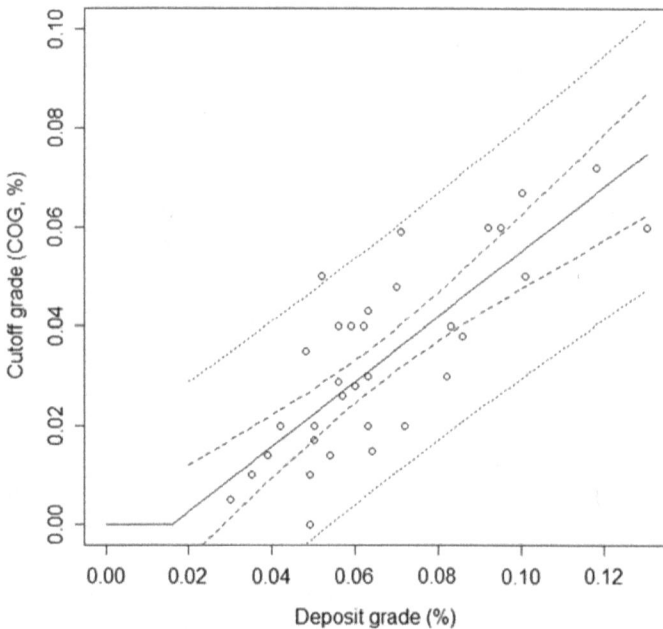

Fig. 20.1 Cutoff grade (COG, %) versus deposit grade (%) plus a fitted linear model and the 95% confidence intervals (dashed lines) and corresponding prediction intervals (dotted lines) for the 34 deposits (Appendix 1)

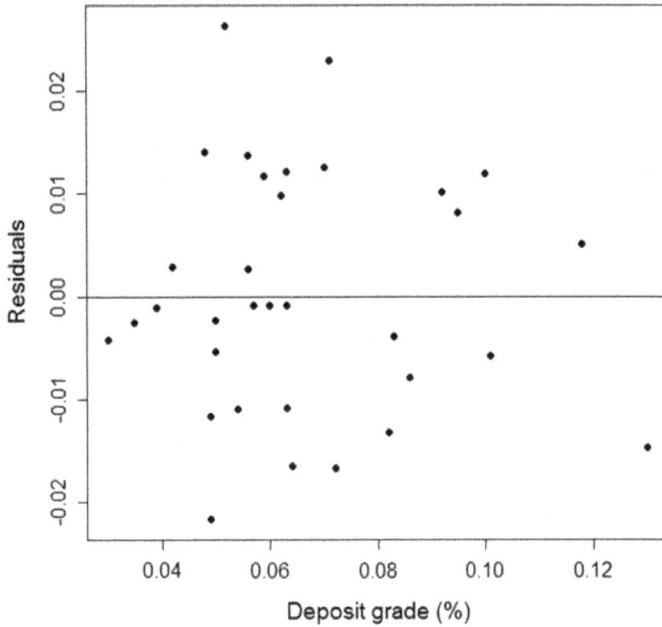

Fig. 20.2 Residuals versus deposit grade for the linear model fit (Fig. 20.1)

The model to fit cutoff grade U as a function of deposit grade D is

$$U = \begin{cases} 0 & 0 \leq D < c \\ \beta_0 + \beta_1 D + \varepsilon & D \geq c \end{cases}$$

where ε is the random error, assumed to be normal $N(0, \sigma^2)$. The constant c is determined from the linear regression fit since the $COG \geq 0$.

The fitted model is:

$$\hat{U} = \begin{cases} 0 & 0 \leq D < c = 0.0159 \\ \beta_0 + \beta_1 D = -0.01042 + 0.6553D & D \geq 0.0159 \end{cases}$$

where $\hat{U}$ is the estimated cutoff grade in percent and D is the deposit grade in percent. The residual standard error is 0.012 on 32 degrees of freedom and the adjusted $R^2 = 0.61$. The model is statistically significant and reasonable for the given data set. The residual plot is shown in Fig. 20.2.

There is no evidence to suggest that the residuals are non-normal. Thus, within the domain of the deposit grade, namely from 0.03 to 0.13, the linear model shown above appears to be appropriate. Predictions outside of this interval will depend on the same linear relationship holding.

20.3   Deposit Growth as a Function of Cutoff Grade

The second model is the fraction of growth as a function of estimated cutoff grade. In this example the growth data (Fig. 20.3) consists of 58 observations from eight deposits (Appendix 2). The inverse S shaped form of the data corresponds to an inverse cumulative distribution function. Therefore, this relationship is modeled as an inverse cumulative distribution function, since the fraction growth is a number between 0 and 1, inclusive. Several models including the gamma, lognormal, normal and their left truncated forms were candidates to fit this data. Of these, the left truncated normal was the best fit by visual inspection and by a nonlinear least squares fit. The form of the left truncated normal probability distribution function is:

$$f_L(x|\Theta) = \frac{f(x|\Theta)}{1 - F(\lambda|\Theta)} \quad x > \lambda$$

where $\Theta' = (\mu, \sigma^2)$ and the left truncation point λ is assumed known. The probability density function for the normal distribution with mean μ and standard deviation σ is:

$$f(x|\mu, \sigma^2) = \frac{e^{-(x-\mu)^2/2}}{\sqrt{2\pi}\sigma}$$

The corresponding left truncated cumulative distribution function, cdf, is:

$$F_L(x|\Theta) = \frac{F(x|\Theta) - F(\lambda|\Theta)}{1 - F(\lambda|\Theta)}, \quad x > \lambda$$

The truncated distributions' models used for model fitting are from the package truncdist (r-project.org) by Novomestky and Nadarajah (2012) based upon work by Nadarajah and Kotz (2006).

As Fig. 20.1 shows, there is uncertainty in the COG when estimated from the deposit grade. However, when estimating the left truncated normal cumulative distribution function (cdf), the estimates are conditioned upon the COG being known. A possible alternative is an errors-in- variables approach (Schennach 2004) where both the fraction growth and cutoff grade are considered to be random variables.

The chosen optimization criterion to estimate the fraction growth (Fig. 20.3) is

$$\min(\sum_{i=1}^{n} (F(x_i|\Theta) - \hat{F}(x_i))^2,$$

where x_i is the ith COG and F is the cumulative distribution function. Θ contains the estimated parameters. If F is a normal distribution the parameters would be $\hat{\mu}$ and $\hat{\sigma}$.

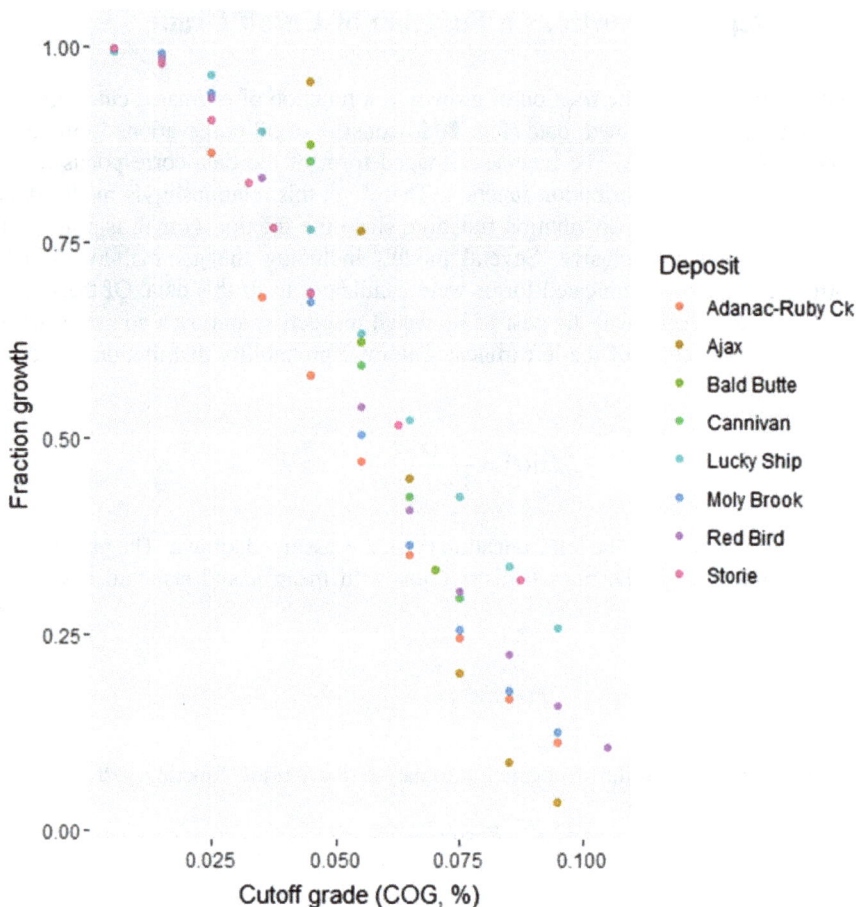

Fig. 20.3 Deposit fraction growth plotted against cutoff grade (COG) in percent for the 8 deposits used in this study

The ith COG is represented by x_i and $\hat{F}(x_i)$. Note that $\hat{F}(x_i) = 1 - \hat{G}(x_i)$ where $\hat{G}(x_i)$ is the fraction growth. The nonlinear least squares package used to estimate the left truncated normal model parameters is nls2 (r-project.org). See Grothendieck (2013). The left truncation point is $\lambda = 0$.

Deposit growth as a function of cutoff grade was modeled for each of the eight deposits (not shown). These results indicate that the data could have been generated from the same population Thus, the observations were pooled and a single model was fit. The reason to fit a cumulative distribution function was twofold. One was that eight deposits were used so the data was not in the form of a stepwise function. The second was that the data were not randomly or systematically spaced across the domain of the empirical distribution. The data, expressed as an empirical distribution function, together with the cumulative left truncated normal distribution fit and confidence intervals, are shown in Fig. 20.4. The results of the least square fit were $\hat{\mu} = 0.0609$ and $\hat{\sigma} = 0.0282$. The residual sum of squares, RSS = 0.3631.

The 95% confidence and prediction intervals for nonlinear estimation are approximate. The confidence interval shown in Fig. 20.4 (dashed lines) is from package propagate, r-project library predictNLS programmed by Spiess (2014) based upon work by Bates and Watts (2007), and others. It uses a second-order Taylor series expansion and Monte Carlo simulation. The second order approximation captures the nonlinearities around $f(x)$. A corresponding algorithm for the prediction interval has not been developed. The prediction interval shown in Fig. 20.4 (dotted lines) is based upon a linear model of the form $H = \alpha_0 + \alpha_1 U + \varepsilon$ where U was the COG. H is a linear estimate of growth. The next step was to estimate the upper and lower prediction intervals for the linear model with $U = 0$, 0.001, 0.002, ..., 0.150. These are vectors **LPIu** and **LPIl** respectively. The upper and lower 95% nonlinear confidence interval vectors estimated above are **CIu** and **CIl** respectively. The differences between the linear prediction intervals and the nonlinear confidence intervals are computed as follows. Let **Lud = LPIu − CIu** and **Lld = CIl − LPIl**. The estimated upper and lower predictions intervals, **UP** and **LP**, for the nonlinear fit (Fig. 20.4) are **UP = CIu + Lud** and **LP = CIi − Lld**. These estimates appear reasonable in the given domain, namely for COG between 0.04 and 0.10.

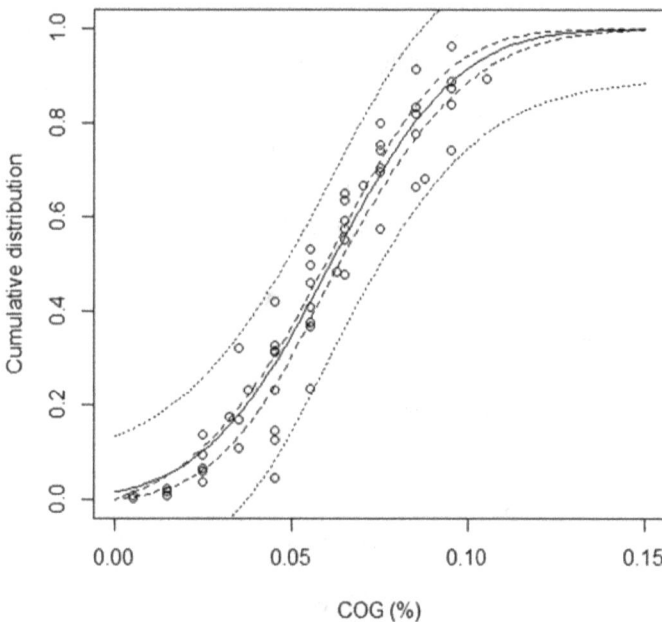

Fig. 20.4 Data fit to a left truncated (at 0) normal distribution is the solid line. The approximate 95% confidence interval is the dashed line. The approximate 95% prediction interval is the dotted line

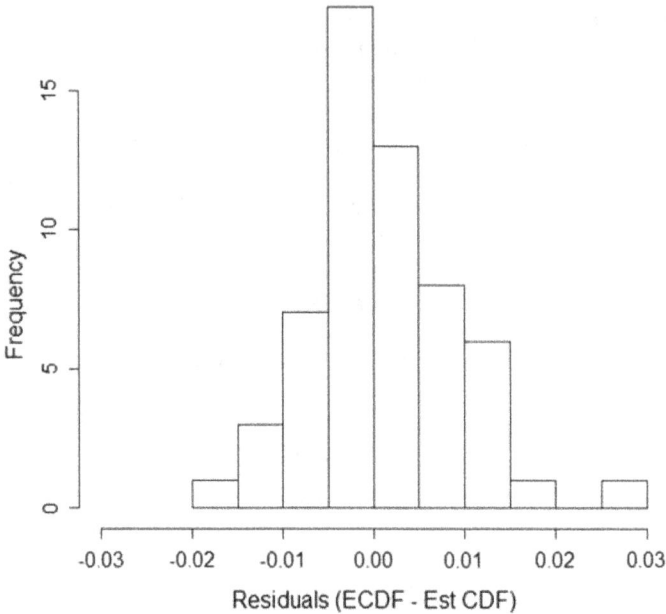

Fig. 20.5 Histogram of residuals for fit to a left truncated normal distribution

A histogram of the residuals, which appear normal, is shown in Fig. 20.5. The truncated normal probability density function corresponding to the cumulative distribution function (Fig. 20.4) and COG data are shown in Fig. 20.6.

Figure 20.7 is like Fig. 20.4 except that the variable plotted on the vertical axis is the fraction growth as opposed to the cumulative distribution. There is no suggestion that the model illustrated in Fig. 20.7 is universal, even for molybdenum deposits. Clearly different deposits may require different models.

20.4　An Example

Suppose the problem is to estimate the fraction growth corresponding to a COG (%) = 0.06 using the model shown in Fig. 20.7. Then, given that the assumed distribution is a truncated normal at zero with estimated model parameters, $\hat{\mu} = 0.0609$ and $\hat{\sigma} = 0.0282$, the results are shown in Table 20.1. The point estimate of fraction growth, namely 0.479, is straightforward to compute. Namely it is:

$$\hat{F}_L(x|\hat{\Theta}) = \frac{F(x|\hat{\Theta}) - F(\lambda|\hat{\Theta})}{1 - \hat{F}(\lambda|\hat{\Theta})}, \quad x > 0, \hat{\Theta}' = (\hat{\mu}, \hat{\sigma}^2)$$

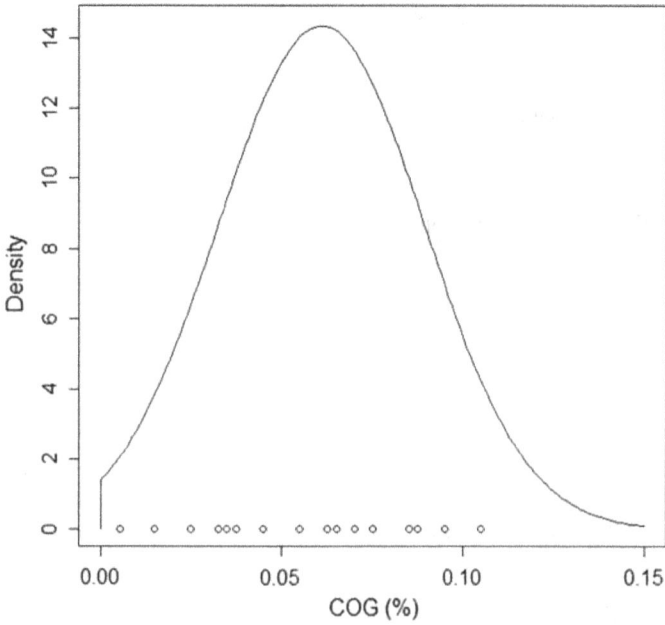

Fig. 20.6 The fitted truncated normal probability density function and COG data (the circles)

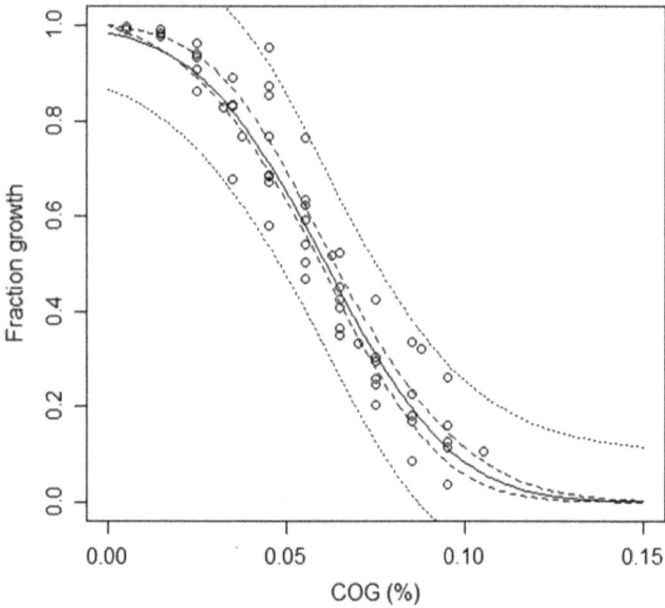

Fig. 20.7 Fraction growth as a function of COG (%) and corresponding fitted values (solid line), 95% confidence interval (dashed line) and 95% prediction interval (dotted line)

Table 20.1 Estimated fraction growth, 95% confidence and prediction intervals for COG (%) = 0.06

		Confidence interval		Prediction interval	
		2.50%	97.50%	2.50%	97.50%
COG (%)	Fraction growth	2.50%	97.50%	2.50%	97.50%
0.06	0.479	0.450	0.507	0.291	0.666

The confidence and prediction intervals are more difficult to compute; however, the R code is available on request from John Schuenemeyer.

20.5   Conclusions

Mineral deposit growth commonly constitutes most unknown resources. The growth considered in this study is due to a progressively lower cutoff grade, which may be unknown. In this study, a statistical model was constructed to model cutoff grade as a function of deposit grade, followed by construction of a model to estimate the fraction growth as a function of cutoff grade. This latter model involves estimation of a truncated normal distribution and second order Taylor series estimates to characterize uncertainty.

Acknowledgements Data used in this chapter represent part of an extensive and ongoing data compilation effort on porphyry molybdenum deposit types. This study evolved over several years through discussions with current and former U.S. Geological Survey employees including Arthur A. Bookstrom, Mark D. Cocker, Robert J. Kamilli, Keith R. Long, Steve Ludington, Barry C. Moring, Greta J. Orris, Ryan D. Taylor, Jay A. Sampson, and Gregory T. Spanski. Eric Seedorf, University of Arizona, Tucson provided his bibliography on porphyry molybdenum deposits, which was of considerable use in this study.

Appendix 1

Porphyry molybdenum data for 34 selected deposits used to model molybdenum cutoff grade as a function of deposit grade.

[Country and state codes: AUQL = Australia, Queensland; CHHN = China; CHNA = China; CNBC = Canada, British Columbia; CNNF, Canada, Newfoundland and Labrador; CNON = Canada, Ontario; CNYT = Canada, Yukon Territory; GRLD = Greenland; MCDA = Macedonia; MNGA = Mongolia; MXCO = Mexico; RUSA = Russia; USAK = USA, Alaska; USID = USA, Idaho; USMT = USA, Montana; USNV = USA, Nevada; USWA = USA, Washington]

Name	ID	Country-State	Mo grade (%)	Deposit size (Mt)	Mo COG (%)
Ada nac-Ruby Creek	101	CNBC	0.042	791	0.020
Adjax-Le Roy	102	CNBC	0.062	552	0.040
Anduramba	103	AUQL	0.054	32	0.014
Bald Butte	106	USMT	0.059	176	0.040
Big Ben	108	USMT	0.092	245	0.060
Buckingham	110	USNV	0.063	1800	0.043
Cannivan Gulch-White Cloud	111	USMT	0.056	327	0.040
Carmi	113	CNBC	0.057	40	0.026
Cave Creek	114	USTX	0.130	28	0.060
Chu	115	CNBC	0.050	673	0.017
Creston	118	MXCO	0.071	215	0.059
Endako	124	CNBC	0.050	1232	0.020
Jiguanshan (Jiganshuan)	130	CHNA	0.095	100	0.060
Joem-Haskin Mountain	131	CNBC	0.101	11	0.050
Kitsault (Updated 11/2015)	132	CNBC	0.070	688	0.048
Lobash	140	RUSA	0.063	365	0.030
Lone Pine	143	CNBC	0.072	179	0.020
Lucky Ship	144	CNBC	0.064	85	0.015
Mac	145	CNBC	0.048	248	0.035
Malmbjerg	148	GRLD	0.118	229	0.072
Moly Brook	151	CNNF	0.049	199	0.010
Mount Hope	152	USNV	0.039	1148	0.014
Mount Tolman	156	USWA	0.056	2200	0.029
Pidgeon-Lateral Lake	163	CNON	0.083	59	0.040
Pine Nut	165	USNV	0.060	181	0.028
Qua rtz Hill	167	USAK	0.082	1310	0.030
Red Bird-Haven Lake	169	CNBC	0.049	201	0.010
Red Mountain	170	CNYT	0.100	187	0.067
Sphinx	178	CNBC	0.035	62	0.010
Storie Molie	180	CNBC	0.049	105	0.000
Sudulica-Mackatica-Kucisnjak-Groznatova Dolina	146	MCDA	0.030	383	0.005
Tangjiaping	183	CHHN	0.063	373	0.020
Thompson Creek	184	USID	0.086	575	0.038
Zuun Mod	195	MNGA	0.052	408	0.050

Appendix 2

Molybdenum data for estimating fraction deposit from cutoff grade; $n = 58$

Deposit name	COG (%)	Fraction growth	Deposit name	COG (%)	Fraction growth
Adanac-Ruby Creek	0.095	0.113	Moly Brook	0.095	0.126
Adanac-Ruby Creek	0.085	0.168	Moly Brook	0.085	0.180
Adanac-Ruby Creek	0.075	0.247	Moly Brook	0.075	0.258
Adanac-Ruby Creek	0.065	0.351	Moly Brook	0.065	0.365
Adanac-Ruby Creek	0.055	0.470	Moly Brook	0.055	0.504
Adanac-Ruby Creek	0.045	0.581	Moly Brook	0.045	0.673
Adanac-Ruby Creek	0.035	0.679	Moly Brook	0.035	0.831
Adanac-Ruby Creek	0.025	0.864	Moly Brook	0.025	0.941
Ajax	0.095	0.037	Moly Brook	0.015	0.991
Ajax	0.085	0.087	Red Bird	0.105	0.107
Ajax	0.075	0.202	Red Bird	0.095	0.160
Ajax	0.065	0.450	Red Bird	0.085	0.226
Ajax	0.055	0.765	Red Bird	0.075	0.305
Ajax	0.045	0.956	Red Bird	0.065	0.409
Bald Butte	0.070	0.333	Red Bird	0.055	0.540
Bald Butte	0.055	0.623	Red Bird	0.045	0.687
Bald Butte	0.045	0.875	Red Bird	0.035	0.833
Cannivan	0.075	0.296	Red Bird	0.025	0.935
Cannivan	0.065	0.426	Red Bird	0.015	0.984
Cannivan	0.055	0.593	Storie	0.088	0.320
Cannivan	0.045	0.854	Storie	0.063	0.518
Lucky Ship	0.095	0.260	Storie	0.045	0.685
Lucky Ship	0.085	0.337	Storie	0.038	0.768
Lucky Ship	0.075	0.426	Storie	0.033	0.827
Lucky Ship	0.065	0.523	Storie	0.025	0.907
Lucky Ship	0.055	0.634	Storie	0.015	0.977
Lucky Ship	0.045	0.767	Storie	0.005	0.998

(continued)

(continued)

Deposit name	COG (%)	Fraction growth	Deposit name	COG (%)	Fraction growth
Lucky Ship	0.035	0.891			
Lucky Ship	0.025	0.963			
Lucky Ship	0.015	0.985			
Lucky Ship	0.005	0.993			

References

Barnes A, Thomas D, Bowell RJ et al (2009) The assessment of the ARD potential for a 'Climax' type porphyry molybdenum deposit in a high Arctic environment: Skellefteå, Sweden, In: 8th international conference on acid rock drainage securing the future (ICARD), p 10

Baudry P (2009) Zuun Mod porphyry molybdenum-copper project, South-Western Mongolia: China: Minarch-Mineconsult Independent technical report dated June 2009 prepared for Erdene Resource Development Corporation (Project No. 3421 M), p 147

Becker LA, Gustin MM, Drielick PE et al (2009) NI 43-101 Technical Report, Creston Project, Pre-feasibility study, Sonora, Mexico: Tucson, Ariz., M3 Engineering & Technology Corporation and Golder Associates, and Reno, Nev., Mine Development Associates, 23 March 2008, to Creston Moly Corp., Vancouver, British Columbia, p 273

British Columbia Ministry of Energy and Mines (2012) Ajax, Le Roy: British Columbia Ministry of Energy and Mines MINFILE No. 103P 223

British Columbia Ministry of Energy and Mines (2014) Huber, Mineral Hill, Butte, Granby, Lone Pine, Independent: British Columbia Ministry of Energy and Mines MINFILE No. 093L 027

British Columbia Ministry of Energy and Mines (2014) Kitsualt, Clary Creek, B.C. Molybdenum, Alice, Lime Creek Lynx, Cariboo, MINFILE Record No. 103P 120

Bates D, Watts D (2007) Nonlinear regression analysis and its applications. Wiley-Interscience

Chen Y, Wang Y (2011) Fluid inclusion study of the Tangjiaping Mo deposit, Dabie Shan, Henan Province: implications for the nature of the porphyry systems of post-collisional tectonic settings. Int Geol Rev 53(5–6):635–655

Committee for Mineral Reserves International Reporting Standards (2006) International reporting template for the reporting of exploration results, mineral resources, and mineral reserves

Drew LJ (1997) Undiscovered petroleum and mineral resources. Assessment and controversy. Plenum Press, New York, pp xiii + 210

Ewert WD, Puritch EJ, Armstrong TJ et al (2008) Technical report and resource estimate on the Carmi molybdenum deposit kettle river property, Greenwood Mining Division, British Columbia: Brampton, Ontario, P&E Mining Consultants Inc., Effective date: 4 August 2008, Signing data: 25 September, 2008; for Hi Ho Silver Resources Inc., Mississauga, Ontario, p 97

General Moly (2012) Mt. Hope. web pages @ http://www.generalmoly.com/properties_mt_hope.php. Accessed 8 Nov 2010

Geological Survey of Finland (2011) Large unexploited deposits in Fennoscandia. web pages @ http://en.gtk.fi/ExplorationFinland/fodd/largeunexpl_060508.htm. Accessed 15 March 2011

Geoscience Australia (2012) Australian Mines Atlas of minerals resources, mines and processing centres, Geoscience Australia. web pages @ http://www.australianminesatlas.gov.au/?site=atlas&tool=search. Accessed 12 June 2012

Grothendieck G (2013) Non-linear regression with brute force. R package nls2. contact: ggrothendieck@gmail.com

Kramer B (2006) Mountain of controversy, tribe's history, future clash in face of mine proposal: SpokesmanReview.com, Web pages @ http://www.spokesmanreview.com/tools/story_pf.asp? ID=116309. Accessed 16 Dec 2010

Long K (2008) Economic life-cycle of porphyry copper mining, In: Spencer JE, Titley SR (eds) Ores and orogenesis; Circum-Pacific tectonics, geologic evolution, and ore deposits: Arizona Geological Society Digest 22, pp 101–110. Ores and orogenesis; Circum-Pacific tectonics, geologic evolution, and ore deposits: Arizona Geological Society Digest 22, pp 101–110

Lowe C, Enkin RJ, Struik LC (2001) Tertiary extension in the central British Columbia intermontane belt: magnetic and paleomagnetic evidence from the Endako region. Can J Earth Sci, vol 38, pp 657–678

Ludington S, Plumlee GS (2009) Climax-type porphyry molybdenum deposits. U.S. Geological Survey Open-File Report 2009–1215, p 16

Mercator Minerals (2011) Molybrook: mercator minerals. http://www.mercatorminerals.com/s/ OtherProjects.asp. Accessed 8 May 2013

Mindat.org (1992) Buckingham mine (Hardy mine; Bentley mine; O'Leary mine), Battle Mountain District, Lander Co., Nevada, USA: Mindat.org. web page @ http://www.mindat. org/loc-60116.html. Accessed 29 Oct 2010

Mindat.org (2011) Mačkatica ore field, Čemernik Mts., Serbia; Mindat.org. web page @ http:// www.mindat.org/loc-40789.html. Accessed 21 Jan 2011

Nadarajah S, Kotz S (2006) R programs for computing truncated distributions. J Stat Softw http:// www.jstatsoft.org/v16/c0. Accessed 16 Aug 2006

Nanika Resources Inc (2012) Management discussion and analysis for the six months ended 31 March 2012. Form 51-102F1, Nanika Resources Inc., filed 30 May 2012, p 12

Northern Miner (2010) Canadian & American MINESCAN Folio, 2009–2010, CD ROM

Novomestky F, Nadarajah S (2012) R-project, package truncdist, updated 20 Feb 2015. contact: fnovomes@poly.edu

Raw Materials Group (2011) Sphinx molybdenum deposit, Canada. Raw Materials Group. Web page @ http://www.rmg.se/RMDEntities/S2/Sphinx_Molybdenum_Deposit_SPHIMO.html. Accessed 26 April 2011

RX Exploration Inc (2010) Drumlummon gold mine, Marysville, Montana, USA. RX Exploration Inc. Web pages @ http://www.pdac.ca/pdac/conv/2009/pdf/core-shack/cs-drumlummon-gold. pdf. Accessed 3 March 2010

Schennach SM (2004) Estimation of nonlinear models with measurement error. Econometrica 72 (1):33–75

Singer DA, Berger VI, Moring BC (2008) Porphyry copper deposits of the world: database and grade and tonnage models, 2008. U.S. Geological Survey Open-File Report 2008-1155, p 45

Smith JL (2009) A study of the Adanac Porphyry Molybdenum Deposit and surrounding placer gold mineralization in Northwest British Columbia with a comparison to porphyry molybdenum deposits in the North American Cordillera and Igneous geochemistry of the Western United States, University of Nevada, unpublished Master's Thesis, p 198

Spiess AN (2014) Package propagate. contact: a.spiess@uke.uni-hamburg.de

Taylor RD, Hammarstrom J, Piatak NM, Seal RR II (2012) Arc-related porphyry molybdenum deposit model, chap. D of Mineral deposit models for resource assessment. U.S. Geological Survey Scientific Investigations Report 2010–5070–D, p 64

Thompson Creek Metals Company Inc (2011) Annual report pursuant to section 13 or 15(d) of the Securities Exchange Act of 1934, for the fiscal year ended 31 December 2011: United States Securities and Exchange Commission Form 10-K, p 140. Web Pages @ http://www. thompsoncreekmetals.com/s/Annual_Report.asp?DateRange=2011/01/01...2011/12/31. Accessed 4 June 2012

TTM Resources Inc (2009) TTM Chu Molybdenum Project accepted by BC and Canadian Environmental Assessment Agencies & SGS Lakefield preliminary metallurgical results: TTM Resources Inc. Press Release dated 4 May 2009. Web pages @ http://ttmresources.ca/english/ wp-content/documents/09-05-04.pdf. Accessed 26 March 2010

U.S. Geological Survey (2011) Buckingham molybdenum deposit: mineral resource data system (MRDS) deposit ID 10155557. Web pages @ http://tin.er.usgs.gov/mrds/show-mrds.php?dep_id=10310305. Accessed 17 Dec 2010

Wu H, Zhang L, Wan B et al (2011) Re-Os and 40Ar/39Ar ages of the Jiguanshan porphyry Mo deposit Xilamulun metallogenic belt, NE China, and constraints on mineralization events. Miner Deposita 46:171–185

Yukon Geological Survey (2005) Red Mountain: Yukon MINFILE No. 105C 009. Web pages @ http://data.geology.gov.yk.ca/Occurrence/12735. Accessed 1 April 2011

Zientek ML, Hammarstrom JM (2014) Mineral resource assessment methods and procedures used in the global mineral resource assessment. In: Zientek ML, Bliss J, Broughton DW et al (eds) Sediment-hosted stratabound copper assessment of the Neoproterozoic Roan Group, Central African Copperbelt, Katanga Basin, Democratic Republic of the Congo and Zambia: U.S. Geological Survey Scientific Investigations Report 2010–5090–T, Appendix A, pp 54–64

21

Mineral Resources and Mineral Resource Estimation

Guocheng Pan

Abstract Mineral target selection has been an important research subject for geoscientists around the world in the past three decades. Significant progress has been made in development of mathematical techniques and estimation methodologies for mineral mapping and resource assessment. Integration of multiple data sets, either by experts or statistical methods, has become a common practice in estimation of mineral potentials. However, real effect of these methodologies is at best very limited in terms of uses for government macro policy making, resource management, and mineral exploration in commercial sectors. Several major problems in data integration remain to be solved in order to achieve significant improvement in the effect of resource estimation. Geoscience map patterns are used for decision-making for mineral target selections. The optimal data integration methods proposed so far can be effectively applied by using GIS technologies. The output of these methods is a prognostic map that indicates where hidden ore bodies may occur. Issues related to randomness of mineral endowment, intrinsic statistical relations, exceptionalness of ore, intrinsic geological units, and economic translation and truncation, are addressed in this chapter. Moreover, a number of specific important technical issues in information synthesis are also identified, including information enhancement, spatial continuity, data integration and target delineation. Finally, a new concept of dynamic control areas is proposed for future development of quantification of mineral resources.

21.1 Introduction

Instead of elaboration of new techniques, this chapter focuses on fundamental aspects in mineral resources assessment (Pan et al. 1992). Some of the critical issues are reconsidered here with respect to new understanding of basic geo-relations

G. Pan (✉)
China Hanking Holdings, 227 Qingnian Avenue, Hanking Tower 22nd Floor, Shenyang 110016, Liaoning, People's Republic of China
e-mail: gpan100@yahoo.com; pangc@hanking.com

between resource descriptors and geological processes. Various multivariate models and techniques have been used over the past two decades to relate geological variables to some aspects of mineral occurrence or deposits. Conventional objective methods for mineral resource assessment have estimated either mineral endowment or discoverable mineral resources of a particular type of deposit in a region. The mineral endowment of a region usually refers to that quantity of mineral in accumulations meeting specified physical characteristics, such as grade, size, and depth. A multivariate endowment model is essentially characterized by a particular information extraction strategy for the so-called optimum combination of those geological features most related to spatial variations of endowment (Pan and Harris 1991). Most of these models estimate mineral resources based upon the principle of analogy, i.e., the resources in a study region are estimated by a model that is established on a control area by assuming different regions with similar geological environments have similar endowment (Pan and Harris 1991; Harris 1984; Harris and Pan 1991; Pan and Harris 2000; Agterberg 1981, 2014).

Most of these models have employed as information reference a grid of regularly spaced cells (inter-grid areas) and have dealt in one way or another with either mineral favorability, probability, mineral wealth or density of mineral occurrence (deposit). Of special interest have been those models that describe uncertainty about these estimates, such as the probability for occurrence of mineral deposits within a cell. These studies seem to have been a necessary step in the evolution of the science of mineral resources prediction, because geologists in general have been slow to adapt quantitative methods, and even reluctant to substitute objective and quantitative analysis for all or part of subjective analysis. Thus, there was a need to demonstrate quantitative methods that could be used to estimate undiscovered mineral resources. However, to some extent, this reluctance represented the dissatisfaction by geologists for the at-best low, and sometimes trivial, level of geoscience information captured by the quantitative variables and related to mineral occurrence by the multivariate models. Simply stated, mineral resource estimates by quantitative and objective methods will not improve significantly until more geoscience information is related in more appropriate ways to the various descriptors of mineral resources.

Supplying worldwide demand of metallic raw materials throughout the rest of this century may require multiple times the amount of metals contained in known ore deposits (Patiño Douce 2016a, b). Sustainability of resource supply is a key task for scientific mineral assessments. The concept of mineral resource is many faceted, including physical and chemical properties of mineral deposits, as they occur naturally in the earth's crust and economic properties created by man's socio–technical production system and the demands for mineral materials derived there from. The discussion presented here focuses upon several aspects of mineral resources that are fundamental considerations in the effective information synthesis for mineral resource estimation: randomness of mineral endowment, basic statistical relations, scarceness, geological foundations, economic truncation and translation, and spatial continuity. Some major issues in quantitative mineral resource estimation are addressed,

including information enhancement, information synthesis, as well as target identification. Information synthesis is a central task in both mineral exploration and
resource estimation.

21.2   Randomness of Mineral Endowment

Most of the past and current studies on mineral resource estimation have been
constructed and applied on the basis of a common assumption that mineral
endowment descriptors and at least some of the related geologic processes behave
more or less according to certain stochastic rules. The assumption is seldom
challenged, although controversies have continued over four decades, for example,
the types of the stochastic laws that govern the true distributions of geochemical
element concentrations (Harris 1984; Vistelius 1960; Brinck 1972). This seems to
indicate that the assumption that some geological processes are to some extent
stochastic and follow certain stochastic laws has been widely accepted, although it
is premature to assert that all of the geoscience features are stochastic. It is useful to
examine this notion before investigating specific stochastic laws for particular
geologic events, the use of statistical models to estimate mineral resources, and
probabilistic descriptions of resource descriptors.

In his famous 'Ideal Granite Model', Vistelius (1972) showed that the crystal-
lization of minerals, such as potassium feldspar, quartz, as well as plagioclase
contained in the 'ideal granite' can be modeled by some stochastic functions that
vary in space and time. It has been proved mathematically that there is a
three-dimensional 'packing of particles' such that the three mutually perpendicular
directions can be described according to the Markov property in each direction with
identical transition probability matrices in the three directions (Vistelius and Harbaugh 1980). Another example due to Vistelius is his gravitational stratification
package model (Vistelius 1981). In the study of red beds of the Cheleken Peninsula, under certain assumptions, Vistelius showed that the sequence of red beds
with two distinct states, S (arenaceous beds) and A (argillaceous beds), can be
treated as a homogenous reversible Markov chain of second order, with the partial
transition through A being first order Markov and the partial transitions through S
being second-order Markov.

Sedimentary sequences have been regarded generally as some types of cyclic
processes which are associated with certain Markov properties (Schwarzacher
1969; Hattori 1976; Pan 1987; Kantsel 1967; Pan and Porterfield 1995). Pan (1987)
demonstrated that many sedimentary sections can be treated as homogeneous
stochastic processes if no significant depositional discontinuities or structural
unconformities occur in the sequences and that homogeneous sedimentary processes can be decomposed uniquely into the sum of independent reversible and
unidirectional stochastic flows.

The process of ore deposition was closely examined by Kantsel (1967) based
upon the function of metal distribution in ores. The process of hydrothermal

mineralization during a single stage can be treated as a continuous stationary process of the Markov type. The resulting concentration of metal can be represented by a distribution function, the most important characteristic reflecting speed of the mineralization process. Stochastic modeling methods and uncertainty quantification are important tools for gaining insight into the geological variability of subsurface structures and formation of mineral deposits (Wang et al. 2017). Modeling of 3D geological processes helps reveal hidden information on the variability of controlling factors, which defines likelihood of occurrence of mineralization processes.

These contributions are informative about some fundamental and crucial controversial issues regarding the application of stochastic models to mineral exploration, although some concerns cannot be satisfactorily resolved without more research. A partial conclusion drawn from these preliminary works should be that *at least under certain conditions some of the geologic or earth processes can be modeled by stochastic laws.* However, it would be incorrect to associate the earth processes with the stochastic laws through one to one relations, since the random properties of geologic events generally are space and time dependent.

21.3   Fundamental Geo-process Relations

Observations on geologic features in certain spatial and temporal settings are the outcomes of a sequence of geologic processes superimposed during crustal evolution and initiated by inner energies of the earth, biosphere, hydrosphere, atmosphere, as well as other universal forces. Conceptually, there should be two levels of cause–effect relations among the geologic events, crustal evolution and initial forces, that created the earth. The earth commonly represents the entity of earth processes, e.g., crustal movement, magmatic intrusion, migration of ore-bearing fluids, erosions, etc., while geologic entities, such as lithologic phases, hydrothermal alterations, geologic structures, ore deposits, etc., are outcomes of the processes. Let $o_1, o_2, \ldots, o_k$ denote the k initial forces, $f_1, f_2, \ldots, f_p$ the p earth processes, and $z_1, z_2, \ldots, z_m$ the m geological features, including resource descriptors. Then, the cause–effect relations may be conceptualized as follows:

$$f_j = g_j(o_1, o_2, \ldots, o_k), \quad j = 1, 2, \ldots, p, \tag{21.1a}$$

$$z_i = h_i(f_1, f_2, \ldots, f_p), \quad i = 1, 2, \ldots, m. \tag{21.1b}$$

The conceptual model (21.1a, 21.1b) implies that the original forces are direct causes of the crustal evolution represented by a series of geologic processes which in turn are the direct causes of the geologic features (outcomes). Since some of these geologic features are resource descriptors, such as number of deposits, quantity of endowment, etc., relation (21.1a, 21.1b) states that a mineral deposit is the result of a sequence of superimposed geologic processes. The functions g_j's and

h_i's may be assumed to be random, provided that the original causes or geologic processes are considered to be stochastic.

A relevant question in statistical estimation of resources concerns basic statistical models useful for describing inherent relations between the geodata and resource descriptors given that geoscience information is stochastic. One should keep in mind the basic cause-effect relations (21.1a, 21.1b) and that these cause-effect relations do not imply any cause-effect between the resource descriptors and other geological features, although syngenetic or parallel relations do exist because both of these are outcomes of some common earth processes. For example, both argillic alteration and copper mineralization result from the same process of magmatic intrusion. Since the current knowledge on the original causes is very limited, it is not realistic to discover relations g_j's in (21.1a, 21.1b). Assuming that the random portions of the earth's processes can be isolated from the deterministic part, the following two sets of auxiliary relations should be essential:

$$r_l = \varphi_l(f_1, f_2, \ldots, f_p) + v_l, \quad l = 1, 2, \ldots, d, \tag{21.2a}$$

$$z_i = \psi_i(f_1, f_2, \ldots, f_p) + e_i, \quad i = 1, 2, \ldots, m, \tag{21.2b}$$

where r_l's are the resource descriptors, z_i's are other geologic features and v_l's and e_i's are the random errors. However, a further difficulty arises because our knowledge of earth processes is also limited. What one can observe in practice are only the geological features z_j's and maybe part of the resource descriptors. Although there is no direct causal relation between the mineral resource descriptors and other geologic features, their syngenetic and concurrent relations will assure some indirect information from the geologic features about the resources. Hence, the geological processes, and thus the mineral resource descriptors, can be mathematically reconstructed through a reverse functional estimation:

$$f_j = \Psi_j(z_1, z_2, \ldots, z_m) + \omega_j, \quad j = 1, 2, \ldots, p, \tag{21.3a}$$

$$r_l = \Phi_l(f_1, f_2, \ldots, f_p) + \varepsilon_l, \quad l = 1, 2, \ldots, d, \tag{21.3b}$$

where ω_j and ε_l are the random error terms for the geological process and resource descriptor estimates.

Accordingly, if m is much greater than d, a feasible solution for mineral resource estimate may be completed in two steps:

(a) Factor out the f_1, f_2, etc. from relations (21.3a) based upon the known information on the geological features z_i's;
(b) Substitute these estimates of the factors into relations (21.3b) and derive the estimates for the multivariate resource descriptors.

The first step of the manipulation is exactly analogous to factor-type analysis, constructing significant geologic factors (causes) from observable geological features, whereas the second step is regression-type analysis, predicting the resource descriptors (effects) from the geological factors. Consequently, *factor-type and regression-type models should be fundamental multivariate statistical models for quantitative mineral resource estimation, and other relevant statistical methods may be considered as variations and combinations of the two types of method.* That's why the mineral resource descriptors (r) can be statistically estimated through the geological features by the following function:

$$r_l = \Theta_l(z_1, z_2, \ldots, z_m) + \vartheta_l, \quad l = 1, 2, \ldots, d, \tag{21.4}$$

where θ_l the random error. The geological processes are directly created by the initial forces of earth movement, while accumulation of mineral resources is directly resulted from complex interactions of the geological processes. Since the geological processes cannot be directly measured, they must be reconstructed by observable geological features, which can be, in turn, indirectly used to estimate mineral resource descriptors through relation (21.4).

21.4 Scarceness, Rareness, and Exceptionalness

The activities of mineral exploration have been motivated chiefly by economic and social pursuits (Pan et al. 1992). Constantly growing economic and social demands require greater amounts of raw material, including nonrenewable mineral commodities. The conduct of mineral resource exploration is predicated upon the economic return expected from the discovery of new deposits. An increase in the price of a mineral product, which is equivalent to the sum of the marginal rent and marginal extraction cost, indicates that the mineral resource has become scarce. A basic perspective of both geologists and economists is that mineral resources are scarce materials in the crust as they occupy only an insignificant portion of crustal material.

Any major ore deposit may be regarded in principle as an anomalous or rare phenomenon commonly characterized by one or more geological, geochemical, and geophysical features. Consequently, signatures of significant endogenic mineralization are anomalous and exceptional geologic settings (Gorelov 1982). In particular, the formation of a giant deposit is an extremely rare event created by an exceptional combination of earth processes. Rareness of the giant deposits is reflected in both spatial and temporal dimensions. Significant concentrations of a metal usually have a strong affinity or correlation with particular geologic formations and epochs, as well as metallogenic environments. The genesis of giant deposits may be controlled by particular regularities that differ from those controlling the formation of medium and small–size deposits of the same composition. It is also thought that the formation of huge deposits appears to be controlled by a so–called 'ore–controlling structure' (Tomson and Polyakova 1984).

Giant deposits often dominate reserves and production. It is not uncommon for a few supergiant and giant deposits to constitute over 50% of the total metal recoverable under current economic and technological conditions; accordingly, the metal quantity in small size deposits is almost negligible (Laznicka 1983). Conversely, giant deposits typically constitute an insignificant part of the total number of ore deposits.

Thus, the scarcity of a mineral resource is essentially determined by the fact that few giant deposits exist in the crust, but the few that do exist strongly dominate reserves and production. Accordingly, the economic viability of mineral exploration is strongly predicated upon its capability of locating the giant or large mineral deposits through delineating the associated geologically anomalous regions of the crust. Unfortunately, conventional quantitative techniques employed have failed to deal with these important particulars satisfactorily, mainly owing to inability to capture the nature of these exceptional constraints, since these unique deposits rarely exhibit common statistical properties.

The discovery process for some deposit types, e.g., those for which structural, geochemical, alteration, or geophysical signatures are correlated to deposit size or those for which discovery is primarily by drilling and for which size is strongly related to areal extent, is size biased, meaning that large, high-grade deposits tend to be discovered in early stages of the exploration of regions (Chung et al. 1992; Pan and Harris 1991). For such deposit types, the prognostication of exploration outcomes or the estimation of additional resources in undiscovered deposits should take into account the implication of this bias to the tonnages and grades of the undiscovered deposits. However, representing the discovery process of other deposit types, such as vein deposits with great vertical extent or those for which size is only weakly related to exploration anomalies, as size bias sampling may not be appropriate (Stanley 1992). Improvement in locating deposits or in estimating probabilities for their occurrence requires consideration of the exploration effect and the conjunction of improved genetic, tectonic, and other unifying geoscience theories with improved synthesis methods for the effective extraction of information from diverse geodata and improved quantitative models for inference or estimation.

Considering the low concentration of many elements, e.g., 65 ppm for copper, in common crust rock, the presence of a large accumulation (1 to 10 million tons for copper) of metal at concentrations that are mined today requires enrichments by 100 or 1000 s times crustal concentrations and the accumulation of metal from a large amount of common crustal materials into a relatively small volume. Typically, this concentration or accumulation is seen as requiring the successive operations of several enrichment-depletion stages. Since these sub-processes rarely take place at the scale and strength required to form an ore deposit, their joint (sequential) occurrence could be an extremely rare event in both space and time. If each of these processes is assumed to be stochastic, the mineralization process is also stochastic, and thus the formation of ore deposits is deemed to be a rare, random event. To the extent that this assumption is acceptable, the concept of rareness of ore deposits is equivalent to the smallness of the probability for the formation of an economic deposit.

The concept of rareness can be compared to that of exceptionalness described by Gorelov (1982) and the conditional exceptionalness proposed by Pan (1989). Some other terms found in literature carrying similar meanings include atypicality, uniqueness, anomaly, etc. The concept of exceptionalness is important and useful in quantitative mineral exploration. The most general feature of major commercial ore deposits is that the geological structures of their ore fields are exceptional and anomalous compared with those of neighboring areas.

It is noted that scarceness is a term relevant to economic aspects of resources, rareness is more closely associated with statistical (probabilistic) characteristics of mineral occurrences; and exceptionalness should be used in a geological context. More specifically, one would say that ore deposits are probabilistically rare and geologically exceptional, even though the metal derived from them may not be scarce in the economic sense described by Barnett and Morse (Barnett and Morse 1963). These terms are often used to describe the status of mineralization events in a relative sense, but they can be statistically quantified in a rigorous framework.

21.5   Intrinsic Geological Unit

Most traditional resource estimations have been made on the basis of regular inter-grids or cells as the sampling scheme and estimation unit. The "cell" approach is associated with a number of drawbacks. The most significant problem is that geological processes can be reconstructed through observable geoscience features, which are measurable in geological units, not artificial cells. The cell-based measurements tend to distort the intrinsic relations between geological features and mineral resource descriptors. Secondly, quantification of the geological features, spatially correlated and even connected, is difficult to capture essential genetic factors that played key roles of metal enrichment. Finally, the cell-approach easily ignores exceptional conditions for formation of large deposits, which cannot be readily quantified through grids.

21.5.1   IGU Definition

In contrast with a population of cells having multiple attributes, consider a population in which each member consists of a set of genetically related objects, e.g., igneous intrusives and associated altered host rock, and each member is described by fields of the related geologic objects. Here, mineral resource descriptors and geoscience measures are attributes of a group of geoscience fields which in turn are attributes of a set of genetically related geologic bodies. Such a scheme employs a sampling reference for quantification and integration of geoscience information that is *intrinsic* to the deposit type being sought. That is why the Intrinsic Geological Units (IGU) was proposed by Pan (1989) and Harris and Pan (1990).

The concept of intrinsic geological units, formally documented in Pan and Harris (1993), has evolved from the notion of *intrinsic samples* (IS), or *consistent geological area.* The basic ideas behind both notions are identical and a minor difference lies in the procedure for delineation. This concept has some common characteristics with the notion of "geological anomalies" proposed by Zhao (2007) (also see Zhao and Chi 1991), although the procedure of unit delineations differs significantly.

An appropriately delineated IGU is at once a great improvement over the traditional inter-grid area or cell because it represents the joint occurrence of geologic bodies that are genetically related to the mineral resources of interest. Thus, even before geological attributes of the IGU are quantified, the very presence of an IGU implies highly significant geoscience information about geology and mineral resources. In contrast, the cell is simply a geometric reference. Therefore, it is inevitably true that geological attributes of an IGU carry far more geoscience information than do the geological attributes of a cell.

IGUs may be formally defined as *members of a population consisting of sets of genetically related geologic objects that are usually defined by their geofields* (Pan 1989). Each member (IGU) of the population of IGUs constitutes an independent set of geologic objects that are genetically related to each other and to mineral deposits, although generally only some of these members contain ore deposits and mineral resources. Moreover, although a particular member of a population of IGUs contains mineral deposits, it may not be uniformly mineralized everywhere within its volume. In other words, a mineral resource unit generally is a subset of an intrinsic geologic unit.

21.5.2 *Critical Genetic Factor*

Any mineral deposit or mineralization can be considered as an anomalous concentration of one or more elements or their chemical compounds when compared to crustal materials. This anomalous region originated from anomalous genetic processes or their superposition during certain geological epochs. Usually, a genetic model consists of a hierarchy of earth processes—from preconditions to post mineralization preservation—which acted during one or more previous time spans, and as such, these processes are not observable. Instead, the geologist must infer their previous existence and operation using observable indirect evidence, e.g., geologic features, geochemical suites, hydrothermal alteration, aeromagnetic and gravity anomalies, etc.

Since particular genetic processes were initiated and developed under certain specialized circumstances, existence of mineralization, as a significant outcome of the processes, must also be conditional upon these relevant circumstances. In other words, whether an anomalous concentration of a metal exists in a region depends solely upon the existence of certain necessary conditions during crustal evolution. Although there might exist a number of such necessary conditions for a particular genetic process or mineralization, one, or at most a few of them, is referred to as critical. For convenience, this (these) critical or necessary condition(s) is called the *Critical Genetic Factor*(s)

(CGF). The idea of CGF does not rest solely upon one factor being more important or critical than another in the formation of a mineral deposit, because unless all genetic factors are present, there is no mineral deposit or mineral endowment. Criticality, as used here, rests more upon the idea that the CGF arises from few, preferably only one, earth process and that those features formed by that process can be detected reasonably well by conventional sensing technologies, e.g., magnetics, gravity, geochemistry, and geology mapping. If this CGF is not present, the intrinsic geological unit is considered to be absent. For example, for a mineral deposit related to magmatic fluids, the heat source that drives intrusion may be treated as the CGF for identification of the IGUs associated with the deposits of this type. Practically, only a single CGF is necessary for identifying spatial units that are intrinsic for mineral deposits of a single genetic type, but more than one CGF may be necessary when there is more than one genetic type of interest.

An IGU can be further understood to be a member of a population consisting of sets of geologic objects genetically associated with the CGF, each set being a member of the IGU population. Individuals from the population are called *known* IGUs if the related CGF is directly observed, while others are unknown or predicted when the CGF cannot be observed directly, but is inferred to exist because of the presence of geologic fields related to the CGF and to recognition criteria.

21.5.3 Critical Recognition Criteria

The CGF often may be identified as a process, based upon geoscience; conceptually, it may be an abstraction, instead of an observable feature. In order to make the CGF concept workable in practice, a set of special geologic features which give firm evidence of the previous existence and operation of the CGF are established. Such a feature is here termed a *Critical Recognition Criterion* (CRC). Each of these CRCs constitutes a sufficient condition for existence of the CGF. Any spatial location at which one or more CRCs occur is by definition a location within an intrinsic unit.

Although the concepts of CRC make it possible for identification of CGF, the occurrence of CRCs known at the time of application may not represent the entire picture of a CGF. In other words, estimation of the presence of a CGF based upon only CRCs could be biased due to imperfect knowledge on the spatial distribution of CRCs. For example, a CRC might exist underneath the sedimentary cover, even though it is not found by surface geological mapping. This fact dictates that the identification of CRCs beyond surface observation is an important step in the appropriate prediction of the distribution of the CGF. This can be done by establishing statistical relations of each CRC to a set of selected geological, geochemical, and geophysical fields, which provide indirect evidence for the presence of the CGF.

Although the existence of a recognition criterion at a spatial location almost surely indicates that the location is within an IGU, the boundary of the IGU still is unknown. Consider, for example, the outcrop of a Tertiary intrusive assumed to be a CRC. Then, the outcrop area is surely within an IGU, but probably, some of the area around the outcrop also is within the same IGU because of the likelihood that at depth the intrusive extends laterally underneath the surface rocks. Consequently, the boundary of an IGU is usually uncertain. One way of representing such uncertainty is to assign each spatial location a probability for presence of one or more recognition criteria based upon a collection of geological observations at that location.

21.5.4   IGU Delineation

At a known location (with at least one observed CRC), the probability for the CGF should be one or very close to one. This implies that the point is almost surely within an IGU. At an unknown location (with no observed CRCs), all of the CRC probabilities estimated from geoscience fields will provide a measure of the likelihood of the presence of the CGF.

Several methods have been proposed and employed for delineating IGUs. One such example is that which consists of three steps developed by Pan and Harris (1993). The method delineates IGUs by estimating and combining probabilities of CRCs. Another example is given by Pan (1989) and Harris and Pan (Harris and Pan 1991) based on the union of marginal field anomalies. As discussed, the presence of a CRC gives evidence for the existence of an IGU; delineation of the boundary of the IGU is made by resolution of the geoscience fields associated with the CRCs. In this approach, the key step is to establish a procedure to identify the anomalies in terms of CRCs for each geosciences field. These anomalies (called marginal anomalies) are then combined into one anomaly through spatial union. This is similar to the concept of using the maximum CRC probability to represent the probability for CGF.

As we know, genetic theories are most useful for grass-roots exploration or reconnaissance programs, where deposit information is not abundant. Without the guidance of genetic models, it is unsafe to select an area for a massive investment. Hence, the concept of IGU is most useful for regional mineral exploration, because it provides a quantitative framework for delineation of those areas having the conditions necessary for the presence of deposit. In large-scale exploration, such as deposit or district scale, the methodology of IGU is still useful if detailed aspects of deposit genetic models can be specified. With abundant occurrence information, it is possible to extract genetic factors as necessary conditions for the localization of deposit. However, in most cases, this detailed information is not available or not in a usable form. In general, a mining district is already a known IGU defined by broad genetic models. Unless refined genetic models are available, IGU will not provide additional power to identify areas for the potentials of deposit or district scale.

21.5.5   Relations Between IGU and Mineral Target

As discussed, CGF serves as the necessary condition for presence of an IGU, but it is not a sufficient condition for the boundary definition of the IGU. The purpose of IGU proposal is to improve methodology of target identification and delineation, which, in turn, improves the effect of mineral resource assessment. The IGU theory creates a new platform on which new approach to mineral target identification can be constructed. A critical question to ask would be what is the relation between IGU and mineral targets?

Theoretically, an IGU is a necessary condition for presence of mineralization of interest. The concept of IGU provides a precursor to the identification of mineralization or deposits. However, presence of an IGU does not necessarily serve as sufficient conditions to the presence of mineralization or deposit. Presence of an IGU is a necessary condition of presence of mineral target. In general, an IGU is much broader in areal or volumetric extents than a mineral target. Mineral targets are defined in the IGU areas where additional necessary and even sufficient conditions are observable or inferable from maps or data collected from various sensing or engineering technologies. Instead of using an inter-grid sampling scheme, the framework of IGU provides a more practical and useful approach for extraction of sufficient conditions for identification of mineralization events through reconstruction of geological processes that resulted in the occurrence of mineralization.

For mineral resources appraisal, the concept of IGU establishes a theoretical base for definitions of necessary and sufficient conditions of mineralization or deposit. It has radically changed the conventional methodology for estimation of mineral potentials. The relationships of IGU, target, occurrence, and deposit are depicted as follows:

$$Deposit \subseteq Mineral\,Target \subseteq IGU \subseteq Working\,Area$$

Clearly, an IGU is not a mineral target, but a mineral target must be enclosed in an existing IGU. Similarly, a mineral target is not a deposit, but a deposit must be localized inside an existing mineral target. Therefore, identification and delineation of IGUs is a necessary step for definition of mineral targets. This new approach will play a revolutionary role in improvement of mineral resources assessment.

21.6   Economic Truncation and Translation

Mineral deposit is not a purely geological concept when it is linked to resources and reserves. The effects of economic truncation and translation on mineral deposits have been recognized several decades ago, and a thorough discussion of these has been given by Harris (1984). These phenomena reflect an important fact that mineral resources generally are a dynamic function of relevant economic and technologic constraints, including price of product and costs associated with various production phases, such as mining, milling, smelting, as well as refining. Available data on mineral deposits generally are truncated by a cost surface which is defined in terms of physical features of the deposits and technological states. In other words, the collection of mineral deposits reported reflects only the truncated fraction of the entire population of mineral deposits. Thus, use of these data directly and unavoidably results in biased estimates of mineral resources, as the characteristics of the resource distribution derived from the partial data set only are a distorted representation of deposits as they occur in nature.

Translation refers to the fact that commonly reported deposit grades and tonnages are for ore reserves and that these tonnages and grades generally differ from those for the total mineralized material for the deposit as a geologic phenomenon. For deposit types having great lateral or vertical gradation in mineralization, economic rents may lead to the selection of a cutoff grade that leaves part of the deposit in the ground. When this is the case, reported ore tonnage is smaller than deposit tonnage and average grade is higher than deposit average grade.

The importance of translation as a distortion varies with the mineral commodity and the maturity of the exploration activity. In general, the greater variation of the grade within a deposit (intra deposit grade variance), the stronger the translation effect, and vice versa. For those deposit types having sharp boundaries or a uniform grade distribution, the translation effect may be negligible. For some deposit types, it is also true that the longer the deposit has been mined, the greater the reserve additions and the more representative the revised ore tonnage and grade data are of the geologic deposit.

The truncation and translation effects are related to some degree when production costs are strongly influenced by ore tonnage and ore average grade, provided that intra deposit grade variation and the spatial distribution of grades permit the effective use of cutoff-average grade relations to maximize the net present value of economic rents. However, translation occurs mainly in mine development and subsequent mining, while truncation reflects both exploration and mining. Conversion of resources to reserves involves using cutoffs for grades that define boundaries of ore economic portions in the deposits. This procedure involves both translation and truncation.

In order to resolve these difficulties, Harris (1984) suggested a possible remedy: treating the truncation effect requires first identifying the truncation relationship, and second the explicit consideration of this relationship in the estimation of parameters, one of which is the correlation of deposit tonnage with grade. Although several attempts have been made to mitigate the difficulty in practical studies by employing more sophisticated mathematical methods in mineral endowment estimation, the problem remains to be explored further, as estimation of the cost relation is still based on the truncated data. Thus, the cost relation must be reconstructed from a truncated surface before estimation is carried out.

The importance of truncation and translation effects on a quantitative estimate of mineral resources depends to some degree upon the means of estimation and upon the objective of the estimation. For example, when estimation is to be done using analogue or control regions and the objective is to estimate the magnitude of resources for price, cost, and technology similar to those of the analogue regions, the effect of truncation and translation on the estimate may be minor. But, when the objective is to estimate the magnitude of resources for improved exploration and production technology, the effect of truncation and translation upon the estimate may be very significant.

21.7 Information Synthesis

The geologist's view of an ore deposit may differ from that of the economist. Economists tend to consider an ore deposit as being a continuous geologic phenomenon that is discretized by applying a set of economic regularities, while geologists tend to perceive a deposit to be a discrete geologic phenomenon with anomalous concentration of one or more valuable elements (Agterberg 1981). Physical mechanisms of ore genesis suggest that the continuity of ore concentration is meaningful mainly in a relative sense. A high magnitude of element concentration in host rocks often contrasts sharply with concentrations in surrounding wall rocks. This perspective may be partially illustrated by the DeWijs' scheme of element enrichment in a deposit, which was extended by Brinck (1972) to describe element concentrations within the crust. Another well-known hypothesis is Skinner's bimodal proposition of element distribution which asserts that a gap exists between the grades of mineralized rock and the grades of common crustal material (Skinner 1976).

21.7.1 Spatial Continuity

Although the continuity of the statistical distribution of grades seems to differ conceptually from that of spatial and temporal distributions, they are in fact closely related. For example, if the proposition is accepted that the grades of an element are continuously distributed in space and time, the continuity of the statistical distribution of these grades can be automatically invoked in certain environments, and vice versa. This assertion may be explained by the requirement that samples must be taken in a uniform and regular manner from the population of interest.

Metallogenic and tectonic studies depict elements to be concentrated in geologic terrains of different scales, such as ore shoot, ore body, ore district, ore belt, ore province, etc. (Laznicka 1983). This hierarchical structure of ore formation seems to indicate that continuity exists within each of these scales, while discreteness of ore concentrations can be seen between these different scales. For instance, an ore district may be viewed as a continuously anomalous region within an ore belt, but the individual deposits included in that same district are discrete geological phenomena. This perspective carries strong implications as to sampling procedures and the organization of data for the estimation of mineral potentials.

Thus, a specific mineral exploration project focused upon the ore deposits of certain valuable elements formed and confined in a particular dimensional scale requires an appropriate sampling scheme of that same scale. For example, a new ore body developed within a deposit may be considered as mineral potential at the deposit scale, while a new ore deposit discovered in a district is regarded as mineral potential at a district scale. When estimation is aimed at predicting the mineral potentials at the district scale, the sampling scheme must accommodate the geological and mineral continuity at the corresponding hierarchical level. The match in scale is a prerequisite in mineral resource estimations.

21.7.2 *Information Enhancement*

Although in one sense considerable progress is apparent in the use of quantitative techniques for mineral exploration and resource estimation since the early work in the 1950s and 1960s (Allais 1957; Harris 1965), much less success has been made in creating estimates that are or have been used in mineral exploration and mineral policy decisions. Even though quantitative estimation of local/drilling targets may require the detailed quantitative characterization of favorable geological, geochemical, and geophysical information, many explorationists still favor subjective and qualitative methods for the integration of geodata. Concurrent with these applications, mathematical methods were designed and demonstrated, but few were adopted. Perhaps, this is a natural evolution of the science of quantitative mineral exploration in terms of data integration, because geologists in general have been slow to adopt quantitative techniques. However, this reluctance is at least partly related to ineffective integration of geodata and insufficient extraction of geoscience information by quantitative models. Mineral resources cannot be satisfactorily estimated until more geoscience information is related by improved methods to mineral occurrence. Major difficulties that have hindered further development have been far from fully attacked, and some of them are even completely ignored.

A common practice in quantitative mineral exploration is to collect all relevant geoscience data available in the study region, including numerical observations, digitized maps, and remotely sensed images. These data are then compiled, digitized, resorted, and formatted in a readily manageable data base. Each record is usually stored as a row, while each geologic attribute occupies a column. In standard statistical terms, each record in a data base is called a sample and each attribute is referred to as a variable. A sample in mineral exploration can be a spatial point or a one-, two-, or three-dimensional block. Most data in regional mineral exploration are interpreted in two dimensional areas.

Sampling schemes are considered to be an important factor in data interpretation and target identification. A viable sampling scheme should be able to cope with the hierarchical structures of mineralization or ore concentration. Mineralized geological bodies in different hierarchical scales correspond to different domains in space and time, which are generally defined by particular tectonic settings and geological formations. Statistically, samples should be randomly taken in the population of mineralized and non-mineralized geological blocks of the same scale. Furthermore, spatial characterization of geological features is another criterion for reasonable representation of the resource variability. A reliable sampling scheme should also result in a sample distribution which portrays closely the 'true' population distribution of geological and mineralized bodies. Our experience has shown that quantities measured on the basis of equal area cells might lead to distorted probability distributions.

The original data may include geological, geochemical, geophysical, as well as remote sensing information in diverse modes. For example, geological data can be hydrothermal alteration, faults, and lithology, which are typically considered as non-numerical attributes. Geochemical data can be collected from a rock outcrop, stream sample survey, or a soil grid survey. Magnetics data can be obtained from an airborne geophysical survey. It is readily seen that all these types of geodata are diverse not only in terms of sampling methods, but also the presentation of quantities. Different sampling schemes create different data densities, inconsistent spatial locations, disconnectivity, as well as uneven precisions. Different quantity presentations may give rise to even more serious problems in data integration. The most difficult problem is dealing with the correlation of different variables, which is the most critical step in geological information synthesis, especially when some data are non-numerical. The first step in overcoming these difficulties is the quantification and unification of different data sets.

The quantification of non-numerical attributes refers to assignment of a numerical value to each sample location; of course, the numerical value must convey explicit geological information. For example, a binary assignment gives 1 or 0 to the attributes to represent presence or absence. When each data set is 'quantitative', the next step is to enhance geological information of each individual data set before they are compared, correlated, and integrated. As a matter of fact, enhancement of information from original and individual data is the most critical step towards a successful information synthesis for mineral target selection. Unfortunately, geologists traditionally tend to place too much emphasis on the original data and denigrate the importance and necessity of data filtering, cleaning, and enhancing. Conversely, some geomathematicians devote too much attention to processing of data and give too little regard to fundamental characteristics of the original data and the useful information of the data. Original data carry the most genuine information, but they may be 'contaminated' or masked by noise and even distorted due to inadequate sampling or analytical methods.

Filtering and enhancing of useful information is important to remove noise and reveal signals, such as separation of soil geochemical anomalies from background values. Furthermore, one data set may carry information on several geological aspects. Some of these signals are not the major interests and their presence sometimes masks or distracts from the information useful in identifying mineral targets. These signal components are unwanted, even though they are not noise, and should be filtered out, or at least suppressed. However, many filtering, enhancing, and other data processing techniques can easily introduce artifacts or false signatures. For instance, a magnetic anomaly map generated from a short-wavelength filter can exhibit many high-amplitude, single-grid-point anomalies, which are known as the aliasing effect in the geophysical literature. Another example is interpolation which has been commonly used in data interpretation and quantitative mapping. All interpolation algorithms, e.g., minimum curvature and kriging, which can be considered as low pass filters, are notorious in that they tend to produce overly smoothed surfaces and quite often cause a loss of important detailed features. It is our opinion that some applications of quantitative analysis in mineral exploration have either failed to extract the important geoscience information or have created too many artifacts relative to signals; these effects are believed to be among the major reasons underlying the reluctance of geologists to replace qualitative judgment by quantitative analysis.

The above discussion suggests that filtering and enhancing is necessary for geological data interpretation and integration, but care is warranted in the use of enhancing techniques. Also, enhancement of a geological attribute includes identification and description of spatial structural characteristics, which constitute useful information about spatial auto-correlation of the attribute. More specifically, the objective of information enhancement is to maximize the signal relative to noise. By analogy, the best picture of an object taken by a camera requires a correct focus on the object; either too short or too long of a focus will blur the picture. Moreover, one should keep in mind that any enhancement technique cannot create information that is not present; instead, it is only able to reveal important features of the information carried by the attribute. But, without enhancement, some important features may not be identified nor employed in subsequent analyses. Since the amount of information in each attribute is limited, enhancement also is limited. A minimum level is necessary, for an insufficient removal of noise fails to reveal the signals to be extracted and used in subsequent analyses. Generally, the tendency of analysts is to ignore or inadequately remove noise and to over-enhance the signals. Of course, intense enhancement of data that contain noise leads to enhancement of noise as well as the signal and to false patterns and inter-relations with other information.

21.7.3 Data Integration

Synthesis of geoscience information includes the quantification of geological observations, maps, and other geological images; extraction of quantitative variables; statistical preprocessing; filtering and enhancement; estimation of statistical relations among variables; and the combination of different data sets (layers). Clearly, most of the components require some amount of computation which can be performed more efficiently by using a computer. There is an obvious advantage of using a computer when many variations of the same type of analysis are required

(Green 1991) or when important information includes the computer interaction of several large sets of geodata. This additional information helps to reduce uncertainties and ambiguities in geological interpretation and mineral potential estimation. Furthermore, some effective and sophisticated statistical techniques which generally prohibit manual calculations can be readily implemented on a computer.

Mineral exploration generally deals with diverse geological data in various chemical and physical forms. Appropriate information synthesis should reflect the types of information contained in each data set and their geological implications. For example, geochemical information is generally different than geophysical data. Even the same type of data, e.g., geochemical, may require different interpretation when it is obtained through different sampling techniques. For instance, soil geochemical samples are processed in different ways from stream samples. Geophysical data are rich in depth information and are capable of locating blind targets, but the extraction of such information requires appropriate processing and analysis. It is important to note that any data set has its limitations in the diagnosis of geologic favorability for mineralization, and interpretation and information synthesis must recognize these limits. Because of vast differences in geoscience content, precisions of measurement, and scales of reference among diverse geologic data, integration of these data directly cannot constitute their optimum use in mineral exploration unless the data are appropriately preprocessed and unified. Unfortunately, these problems are far less than adequately treated in traditional exploration applications.

Geoscience attributes are usually processed, correlated, and integrated to produce some estimates which characterize the favorability or probability of mineral occurrence. A more comprehensive approach treats each of the various kinds of geoscience information as a field of a particular type, e.g., geochemical fields, magnetic fields, etc. (Harris and Pan 1990, 1991). Mineralization may also be viewed as an ore field. The notion of field enriches useful information about three dimensional characteristics of geological bodies. Such a field is generally more expressive of meaningful geoscience information relevant to mineral resources than are 'man-made' variables, e.g., measurements quantified with regard to an artificial reference, such as a grid.

A major objective of information synthesis is to maximize the extraction of relevant geoscience information in terms of mineral potentials. Geological measurements in mineral exploration are commonly multivariate in terms of either several variables (fields) measured at same sample locations, or different variables measured in different sample locations but in the same study region. In the latter case, synthesis may require an appropriate interpolation of the data before they can be jointly analyzed. When strong correlations exist among the variables, multivariate techniques are necessary to capture the joint information from multiple associations as well as the marginal contributions from individual attributes. A multivariate exploration system sometimes can be decomposed into several less significantly correlated sub systems with smaller dimensions. This partitioning may reduce the complexity of modeling and possibly permit more robust estimates at the expense of decreasing the degrees of freedom in the system.

Optimum combination of different geological data sets (layers) has been a central task in data integration and information synthesis. Agterberg (1989) gives a comprehensive review on some major integration methods developed in recent years. Two major types of models notable in literature include favorability analyses and probability methods. Pan and Harris (1992) propose a weighted canonical correlation method for the estimation of a favorability function. These methods are most suitable for combining continuous geological attributes. Agterberg (1992) provides probabilistic techniques for combining indicator patterns in weights of evidence modeling. Both types of models, however, are deficient in some regards. Favorability methods often carry ambiguities in predicting mineral potentials, whereas evidence combination techniques are subject to strong constraints on the independency of different attributes. Moreover, as an information synthesis method, weight of evidence is simplistic. Another useful combination approach is color (RGB) image composition (Sabins 1987). This type of technique also bears some serious limitations, since most current image processing software systems are only capable of combining a very limited number of 'layers'. Therefore, there is a need for development of more effective combination methods.

Geologic information about mineral occurrence may be roughly grouped into two categories: marginal information contributed from individual variables or fields and joint information contributed from the cross correlations between different variables or fields. The first category of information has been extensively quantified and interpreted in most of the traditional studies on mineral exploration. The second category, however, has been inadequately treated due to complexities and ambiguities. Information from the inter-dependencies of variables can be an important factor in improving the definition of exploration targets, if single exploration variables are ambiguous, noisy, and/or uncertain as to mineral occurrence. Thus, an effective synthesis technique must be able to efficiently quantify and extract the cross-correlation information.

Intuitively, there should exist a combination of variables in multivariate mineral exploration that is sufficient to capture the majority of useful information and at the same time to minimize the effort of manipulation. It is probably incorrect to think that more variables are always preferred. On the contrary, a large set of data almost always contains redundant information which, if not appropriately eliminated, can result in unstable solutions and create noisy estimates. Therefore, another important problem in information synthesis is to select and refine variable sets such that redundant and trivial variables are excluded from consideration.

21.7.4   *Target Delineation*

Mineralization is considered as an anomalous geologic event, because the element is either present in anomalous grades, rare minerals, or in anomalous quantities. The purpose of mineral exploration is to locate economic mineral deposits in such anomalous regions based on direct and most often indirect information (chemical, physical, structural, etc.) and ore genetic theories. Since the direct information, e.g., the concentration of the metal of interest, is usually meager in the early stages of exploration, indirect information (e.g., geological, geophysical, geochemical, remote sensing, etc.) is commonly employed to identify mineral exploration targets. However, the mineralized anomalies, which are distinctive from the surrounding areas in terms of the accumulated metal(s), are typically fuzzy or ambiguous in terms of indirect information. Therefore, ambiguities of information raise an intricate question, i.e., how to 'best' define targets in terms of the maximum inclusion of mineralized rock and exclusion of non-mineralized rock.

Information synthesis produces either a set of processed (enhanced, quantified, integrated) geological, geochemical, geophysical fields, or a single synthesized index characterizing the favorability/probability of mineral occurrence. Based upon the derived grids, maps, or images, all of which are commonly referred to as 'layers', mineral exploration targets can be delineated by overlaying or combining the different layers. Since the synthesized results, however, are generally continuous, some threshold values are necessary to define the boundaries of targets. The traditional approaches to determine the boundaries are generally subjective and tend to introduce too many uncertainties. Obviously, a precise definition of a target is an important exploration problem to be solved.

Delineation of potential mineral targets has been a central task especially in the earlier phases of a mineral exploration program. Target areas have been identified by either subjective or objective analysis. Subjective methods provide opportunity for the maximum use of genetic theories of ore deposits and connect genetic knowledge and geological observations either intuitively by expert geologists or formally by a computer system (Harris and Carrigan 1981; Finch and McCammon 1987; McCammon 1990; Koch and Papacharalampos 1988). Subjective methods have been generally formulated as follows: (i) formulate genetic models, (ii) relate geological observations to genetic processes, and (iii) estimate subjective probabilities of mineral occurrence. Objective (mathematical) methods attempt to maximally use various existing mineral occurrence data and quantified geological variables (Botbol et al. 1978; Chung and Agterberg 1980; Agterberg 1988; McCammon et al. 1983; Singer and Kouda 1988). An objective approach generally consists of three major steps: (i) quantification of geological variables, (ii) estimation of mathematical models, and (iii) extrapolation of the estimated models to identify target areas.

Ore genesis models are crucial in mineral exploration and resource evaluation. Since genetic models of ore deposits are usually constructed on the basis of man's past experience, imagination, and logical inference, they have a natural connection to subjective probability analyses and expert systems, giving such an approach great potential for prediction. However, in practice this approach also is subject to some limitations. First, expert systems are costly to build and to validate; second, the full potential of such systems requires the construction and incorporation of extensive data bases. Without such data bases, estimates may be associated with large uncertainties. Furthermore, genetic models change as knowledge is acquired and geologists often disagree on at least some points of a genetic model; this creates uncertainty about the identification of mineral targets. An obvious advantage of objective methods is the production of relatively robust estimates of mineral potentials by extensively using geological, geochemical, and geophysical data. However, these methods also are deficient in some regards. Without using genetic theories, geoscience information content of the variables may be low and may have poor predicting power, i.e., the estimates often 'at best' reproduce what an expert geologist had recognized.

A useful procedure as a link between the two types of model is outlined as follows. First, based upon genetic theories, identify one or more critical genetic factors which are considered as necessary conditions for ore formation. A mineral deposit is believed to be absent if these genetic factors do not exist. Second, identify a set of recognition criteria that offer 'almost sure' existential evidence for critical genetic factors. Third, estimate the favorabilities or probabilities of occurrence of these recognition criteria based upon multiple geodata sets. Fourth, generate a synthesized favorability or probability measure for the occurrence of critical genetic factor(s) based upon the probabilities estimated in the third step. Finally, potential exploration targets are delineated from the synthesized favorability or probability measure through optimum discretization (Pan and Harris 1990). These targets have been referred to as *intrinsic geological units* with respect to the chosen critical genetic factor(s) (Pan and Harris 1993). These targets are so-called chiefly because they are not delineated directly in terms of mineral deposits, but in terms of the critical genetic factor that is a necessary condition for formation of the mineral deposits.

Upon the completion of target delineation, a decision needs to be made as to which targets should receive high priority to be drilled, as different targets vary in the degrees of favorability of mineral occurrence. This need requires the ranking of the targets in the sequence of drilling plans. Rank estimates may be derived directly from the synthesized fields or index. When a reasonable amount of known information on the metal(s) of interest is available in the study region, the rank estimation can be substantially improved by using a functional relation between the synthesized index and the quantity of metal. Of course, estimation of metal quantities is a difficult task, if not impossible. Such a function for estimation of metal quantities is valid only in a sense of pseudo terms, meaning that the results are meaningful only in a statistical sense. Verification for the results is necessary in later stages of exploration and estimation.

21.8   Prediction with Dynamic Control Samples

Most conventional resource analyses are constructed on the basis of extrapolation of some mathematical relations established in control areas into unknown areas (Pan and Harris 2000). Control areas are commonly employed in geodata integration and for the estimation of mineral resources of a relatively unexplored region. As such estimation is predicated upon the principle of *analogy*, the properties of the estimates are heavily reflective of (1) how good of a geological analogue the control area is of the unexplored region and (2) the economic reference for the estimated resources. When analogue and desired resource estimate is for economic and technologic conditions similar to those that induced the exploration and resource development of the control area, resource estimates produced by a mathematical model estimated on a control area may be unbiased. However, when economic or technologic references for the estimates differ or when the control area is not a good geologic analogue, resource estimates are biased and even totally wrong.

Two different approaches to improvement of estimation by mathematical models estimated on control areas are: (1) use only control areas that are exhaustively explored and (2) extend the mathematical model to include exploration variables (such as those defined in Pan and Harris (1991). Both of these solutions present difficulties however: (1) except for very small regions, there are few regions large enough to make good control areas that are exhaustively explored and (2) information on exploration activities generally is not available for regions large enough to make good control areas. When exploration variables are not explicitly included in the model, identification of an appropriate control area presents a difficult problem, for it must represent an unbiased sample of deposit occurrence and nonoccurrence for the relevant geologic environment. As noted by Chung et al. (1992), to compute unbiased estimates of the probability for deposit occurrence conditional upon a set of geologic attributes, it is necessary to know not only the distribution of various attributes in and near mineral deposits, but also the distribution of the same attributes away from mineral deposits (Cox 1990; Agterberg 2015).

Given the issues presented above, it is necessary to solve the dilemma in the selection of control areas and even method of extrapolations of these control areas into unknown regions. The nature of control areas so far is static, meaning that the control areas are fixed when a mathematical model established from these control areas is extended into unexplored regions. Clearly, this static model is hardly adequate for prediction of a large region with complex variability of geological conditions and mineralization characteristics. In other words, the mathematical model built on a basis of samples collected from a control area is only appropriate when the extrapolated areas have geological conditions identical to those in the control areas. It is deemed invalid when the geological conditions in the estimated areas differ from those in the control areas. Hence, a new concept is proposed here: dynamic control areas, which are characterized as self-improvement of the mathematical models through information gains of extrapolated areas away from the initial control areas. The methodology of dynamic control areas and extrapolation of mathematical models are implemented in three steps as follows:

(1) Select the best explored areas in the working region as the initial control area, from which control samples are collected. On the basis of this sample data, a mathematical model is established through data enhancement, combination of different datasets, and techniques of information synthesis. This mathematical

model is then used as the initial model for extrapolation and prediction of unknown areas in the working region.

(2) Update the mathematical model when the model is used for prediction of an unknown unit based on an expanded control sample through addition of new information of exploration variables and target variables (if any) in the predicted unit. The new mathematical model will be more appropriate to the estimation of unknown units. The decision of model update is predicated upon availability of new known target variable information and variability of geological and mineralization conditions from the initial control areas.

(3) Tests are performed with the updated model with respect to its effect in prediction of known units in the initial control areas and the unknown unit. The updated model would be accepted if the test results are satisfied; otherwise, the models will be reconstructed. Quantification of variability of geological and mineralization conditions in the unknown units plays a key role in the predicting power of the updated mathematical models.

The model update above is in nature an iterative process, which improves predictability of the model in the unknown units. The initial control sample is only used for establishment of the initial mathematical model, which is then updated and optimized as it is extended into the predicted areas through incorporation of new information on the variability of geological environments.

Acknowledgements The author wishes to thank for the guidance of Dr. D. P. Harris for the subject and the useful comments provided by Dr. Frits Agterberg of Geological Survey of Canada and Dr. B. S. Daya Sagar of Indian Statistical Institute-Bangalore Centre.

References

Agterberg FP (1981) Application of image analysis and multivariate analysis to mineral resource appraisal. Econ Geol 76:1016–1031

Agterberg FP (1988) Application of recent developments of regression analysis in regional mineral resource evaluation. In: Chung CF, Fabbri AG, Sinding–Larsen R (eds) Quantitative analysis of mineral and energy resources. D. Reidel Publishing Company, p 1–28

Agterberg FP (1989) Computer programs for mineral exploration. Science 245:76–81

Agterberg FP (1992) Combining indicator patterns in weights of evidence modeling for resource evaluation. Nonrenewable Resour 1:1–16

Agterberg FP (2014) Geomathematics: theoretical foundations, applications and future developments. Quantitative geology and geostatistics, vol 18. Springer, Heidelberg

Agterberg FP (2015) Self-similarity and multiplicative cascade models. J South Afr Inst Min Metall 115:1–11

Allais M (1957) Method of appraising economic prospects of mining exploration over large territories: Algerian Sahara case study. Manag Sci 3:285–347

Barnett HJ, Morse C (1963) Scarcity and growth: in the economics of natural resource availability. Johns Hopkins Press, Baltimore

Brinck JW (1972) Prediction of mineral resources and long–term price trends in the nonferrous metal mining industry: in section 4-mineral deposits. In: Twenty-fourth session international geological congress, Montreal, Ottawa, Canada, p 3–15

Botbol JM, Sinding-Larsen R, McCammon RB, Gott GB (1978) A regionalized multivariate approach to target selection in geochemical exploration. Econ Geol 73:534–546

Chung CF, Agterberg FP (1980) Regression models for estimating mineral resources from geological map data. Math Geol 12:473–488

Chung CF, Jefferson CW, Singer DA (1992) A quantitative link among mineral deposit modeling, geoscience mapping, and exploration-resource assessment. Econ Geol 87:194–197

Cox DP (1990) Development and use of deposit models in the U.S. Geological Survey. In: 8th IAGOD symposium program with abstracts, Ottawa, 12–18 Aug 1990, p A99

Finch WI, McCammon RB (1987) Uranium resource assessment by the Geological Survey. Methodology and plan to update the national resource base. U.S. Geological Survey Circular 994, p 22

Gorelov DA (1982) Quantitative characteristics of geologic anomalies in assessing ore capacity. Intern Geol Rev 24:457–466

Green WR (1991) Exploration with a computer: geoscience data analysis and applications. Pergamon Press, Oxford, p 225

Harris DP (1965) An application of multivariate statistical analysis to mineral exploration. PhD dissertation, The Pennsylvania State Univ., University Park, Pennsylvania, p 261

Harris DP (1984) Mineral resources appraisal–mineral endowment, resources, and potential supply. Concept, methods, and cases. Oxford University Press, New York, p 455

Harris DP, Carrigan FJ (1981) Estimation of uranium endowment by subjective geological analysis—a comparison of methods and estimates for the San Juan Basin, New Mexico. Econ Geol 76:1032–1055

Harris DP, Pan GC (1990) Subdividing consistent geological areas by relative exceptionalness of additional information—methods and case study. Econ Geol 85:1072–1083

Harris DP, Pan GC (1991) Consistent geological areas for epithermal gold-silver deposits in the Walker Lake quadrangle of Nevada and California, delineated by quantitative methods. Econ Geol 86:142–165

Hattori I (1976) Entropy in Markov chains and discrimination of cyclic patterns in lithologic successions. Math Geol 8:477–497

Kantsel AV (1967) Function of metal distribution in ores, as genetic characteristics of mineralization process. Int Geol Rev 9:669–676

Koch GS Jr, Papacharalampos D (1988) GEOVALUATOR, an expert system for resource appraisal: a demonstration prototype for Kalin in Georgia, USA. In: Chung CF (ed) Quantitative analysis of mineral and energy resources. D. Reidel Publication Company, p 513–527

Laznicka P (1983) Giant ore deposits—a quantitative approach. Glob Tectonics Metallogeny 2:41–63

McCammon RB (1990) Prospector III: in statistical applications in the earth sciences. In: Agterberg FP, Bonham-Carter GF (eds) Geological Survey of Canada, Paper 89-9, p 395–404

McCammon RB, Botbol JM, Sinding-Larsen R, Bowen RW (1983) Characteristic analysis— 1981—final program and a possible discovery. Math Geol 15:59–83

Pan GC (1987) A stochastic approach to optimum decomposition of cyclic patterns in sedimentary processes. Math Geol 19:503–521

Pan GC (1989) Concepts and methods of multivariate information synthesis for mineral resources estimation. PhD dissertation, University of Arizona, Tucson, p 302

Pan GC, Harris DP (1990) Three nonparametric techniques for optimum discretization of quantitative geological measurements. Math Geol 22:699–722

Pan GC, Harris DP (1991) Geology-exploration endowment models for simultaneous estimation of discoverable mineral resources and endowment. Math Geol 23:507–540

Pan GC, Harris DP (1992) Estimating a favorability equation for the integration of geodata and selection of mineral exploration target. Math Geol 24:177–202

Pan GC, Harris DP (1993) Delineation of intrinsic geological units. Math Geol 25:9–39

Pan GC, Harris DP (2000) Information synthesis for mineral exploration. Oxford University Press, p 450

Pan GC, Harris DP, Heiner T (1992) Fundamental issues in quantitative estimation of mineral resources. Nonrenewable Resour 1:281–292

Pan GC, Porterfield B (1995) Large-scale mineral potential estimation for blind precious metal ore bodies. Nonrenewable Resour 4:187–207

Patino Douce AE (2016a) Metallic mineral resources in the twenty first century. I. Constraints on future supply. Nat Resour Res 25:71–90

Patino Douce AE (2016b) Metallic mineral resources in the twenty first century. II. Constraints on future supply. Nat Resour Res 25:97–124

Sabins FF Jr (1987) Remote sensing. Principles and interpretation, 2nd edn. W. H. Freeman and Company, New York, p 449

Schwarzacher W (1969) The use of markov chains in the study of sedimentary cycles. Math Geol 1:17–39

Singer DA, Kouda R (1988) Integrating spatial and frequency information in the search for Kuroko deposits of the Hokuroku district, Japan. Econ Geol 83:18–29

Skinner BJ (1976) A second iron age ahead? Am Sci 64:258–269

Stanley M (1992) Statistical trends and discoverability modeling of gold deposits in the Arbitibi greenstone belt. In: the 23rd international symposium of APCOM, Ontario, p 17–28

Tomson IN, Polyakova OP (1984) Mineralogical and geochemical indicators of large ore deposits. Glob Tectonics Metallogeny 2:183–186

Vistelius AB (1960) The skew frequency distributions and the fundamental law of geochemistry. J Geol 68:1–22

Vistelius AB (1972) Ideal granite and its properties. I. The stochastic model. Math Geol 4:89–102

Vistelius AB, Harbaugh JW (1980) Granitic rocks of Yosemite Valley and ideal granite model. Math Geol 12:1–24

Vistelius AB (1981) Gravitational stratification. In: Graid RG, Labovitz ML (eds) Future trends in geomathematics, p 134–158

Wang H, Wellmann J, Li Z, Wang X, Liang R (2017) A segmentation approach for stochastic geological modeling using hidden markov random fields. Math Geosci 49:145–177

Zhao PD (2007) Quantitative mineral prediction and deep mineral exploration. Earth Sci Front 14:1–10

Zhao PD, Chi S (1991) Discussion of geo-anomaly. Earth Sci 16:241–248

22

Mineral Resource Assessment Problems and Types of Errors

Donald A. Singer

Abstract Samples are often taken to test whether they came from a specific population. These tests are performed at some level of significance (α). Even when the hypothesis is correct, we risk rejecting it in α percent of the cases—a Type I error. We also risk accepting it when it is not correct—a Type II error at β probability. In resource assessments much of the work is balancing these two kinds of errors. Remarkable advances in the last 40 years in mathematics, statistics, and computer sciences provide extremely powerful tools to solve many mineral resource problems. It is seldom recognized that perhaps the largest error—a third type—is solving the wrong problem. Most such errors are a result of the mismatch between information provided and information needed. Grade and tonnage or contained models can contain doubly counted deposits reported at different map scales with different names resulting in seriously flawed analyses because the studied population does not represent the target population of mineral resources. Among examples from mineral resource assessments are providing point estimates of quantities of recoverable materials that exist in Earth's crust. What decision is possible with that information? Without conditioning such estimates with grades, mineralogy, remoteness, and their associated uncertainties, costs cannot be considered, and possible availability of the resources to society cannot be evaluated. Examples include confusing mineral occurrences with rare economically desirable deposits. Another example is researching how to find the exposed deposits in an area that is already well explored whereas any undiscovered deposits are likely to be covered. Some ways to avoid some of these type III errors are presented. Errors of solving the wrong mineral resource problem can make a study's value negative.

Keywords Quantitative resource assessment · Decision analysis
Uncertainty · Lognormal

D. A. Singer (✉)
10191 N. Blaney Ave., Cupertino, CA 95014, USA
e-mail: singer.finder@comcast.net

22.1   Introduction

Howard Raiffa (1968, p. 264) noted that statistics students learn the importance of constantly balancing making an error of the first kind (that is, rejecting the null hypothesis when it is true) and an error of the second kind, that is, accepting the null hypothesis when it is false (Fig. 22.1). Raiffa thought it was John Tukey who suggested that practitioners all too often make errors of a third kind: of solving the wrong problem. Raiffa nominated a candidate for the error of the fourth kind: solving the right problem too late. John Tukey believed that it was better to find an approximate answer to the right question, than the exact answer to the wrong question, which can always be made precise. More recently, Mitroff and Silvers (2009) focused mostly on social questions where type III errors occurred and provided many examples of developing good answers to the wrong questions (type III error). Unfortunately concerns of Raiffa, Tukey, Mitroff, Silvers, and others are appropriate for mineral resource assessments. And the concerns should not be limited to classical statistics.

Supply of minerals to society is dependent not only on the total amount of mineral material but also on quality or concentrations, spatial distributions or how scattered the material is, whether it has been found, whether it is remote from infrastructure, and a whole host of other issues such as government policies, production technologies, and market structures. Decision-makers, whether concerned about development of a technology, development of a region, exploration, or land management, are faced with the dilemma of obtaining new information, or allowing or encouraging others to obtain it, and the possible benefits and costs of development if mineral deposits of value are discovered. Decisions about exploration for these resources and their possible development require awareness of various kinds and the import of errors that can be made by analysts in their studies.

A type I error is the rejection of the null hypothesis when it is true. In some fields a type I error is called a false positive. The risk of this error is α, the level of significance. A type II error is the acceptance of the null hypothesis when it is false, also known as a false negative error. The probability of making a Type II error, β,

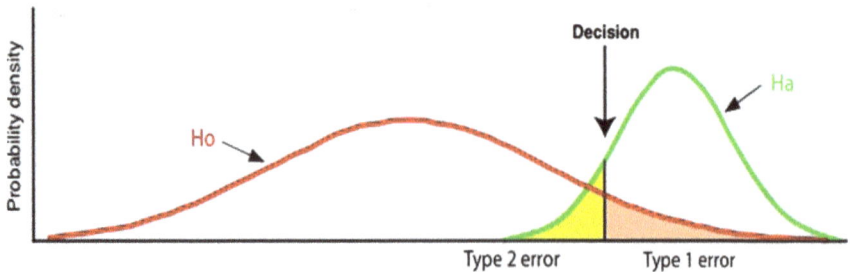

Fig. 22.1 Type I error is the rejection of the null hypothesis (Ho) when it is true. The risk of this is α, the level of significance. Type II error is the acceptance of the null hypothesis when it is false

depends on the alternative value and its distribution. The most important question of the analyst and decision-maker should be: Are we solving the right problem? It is the need to consider this source of error in mineral resource studies that is the focus of this chapter. Common to many of the errors of solving the wrong problem is a mismatch of the studied population and the population that is central to the decisions—this topic is presented first. Next, effects of mismatches of populations to some mineral resource assessments are discussed. Possible ways to avoid some of these type III errors are finally presented.

22.2   Target Population

Type III errors are fundamental and should be considered before errors of types I and II. Type III errors stem from improper definition of the problem and therefore are not strictly a statistical issue, but one of critical thinking. It does no good to minimize the expected costs of type I and type II errors if the wrong problem is being solved. In mineral resource assessments, careless problem definition is the primary source of type III errors. For almost all resource assessment problems, the fundamental sample is the mineral deposit.

The idea of a mineral resource involves both geologic and economic aspects and because knowledge about the earth and future economic conditions is limited, should recognize uncertainty. Mineral deposits are the geologic entities containing resources. Mineral deposits and their contents are the fundamental target populations that are estimated. So what is a mineral deposit? Mineral deposits are defined as mineral occurrences of sufficient size and grade that they might, under favorable circumstances, be economic.

A map of some volcanogenic massive sulfide deposits from Northern Japan is used to clarify our understanding of what is a deposit (Fig. 22.2). From this plot one can see that some of the deposits are just a few meters apart from each other. Grade and tonnages are available for 23 of these named deposits from the western part of the Hokuroku district, Japan (Ohmoto and Takahashi 1983). It is important that if a different map scale were used, this part of the district might have three or four named deposits with grades and tonnages. This well-studied district has more detailed maps than many other volcanogenic massive sulfide districts around the world. If one gathered all available data on the names and grades and tonnages of volcanogenic massive sulfide deposits and built grade and tonnage or contained metal models, the models would contain metals double counted from deposits reported at different map scales and from the same deposits with different names due to grouping. To have a consistent sampling unit that can be applied in statistical analysis and in assessments of undiscovered deposits it is necessary to have spatial rules to help define a deposit. In addition, mine names and deposit names do not always match, mine names sometimes change over time, and district and deposits can be reported with different names and numbers. For example, careless data gathering might contain the grades and tonnage of the total Sudbury Ni-Cu District

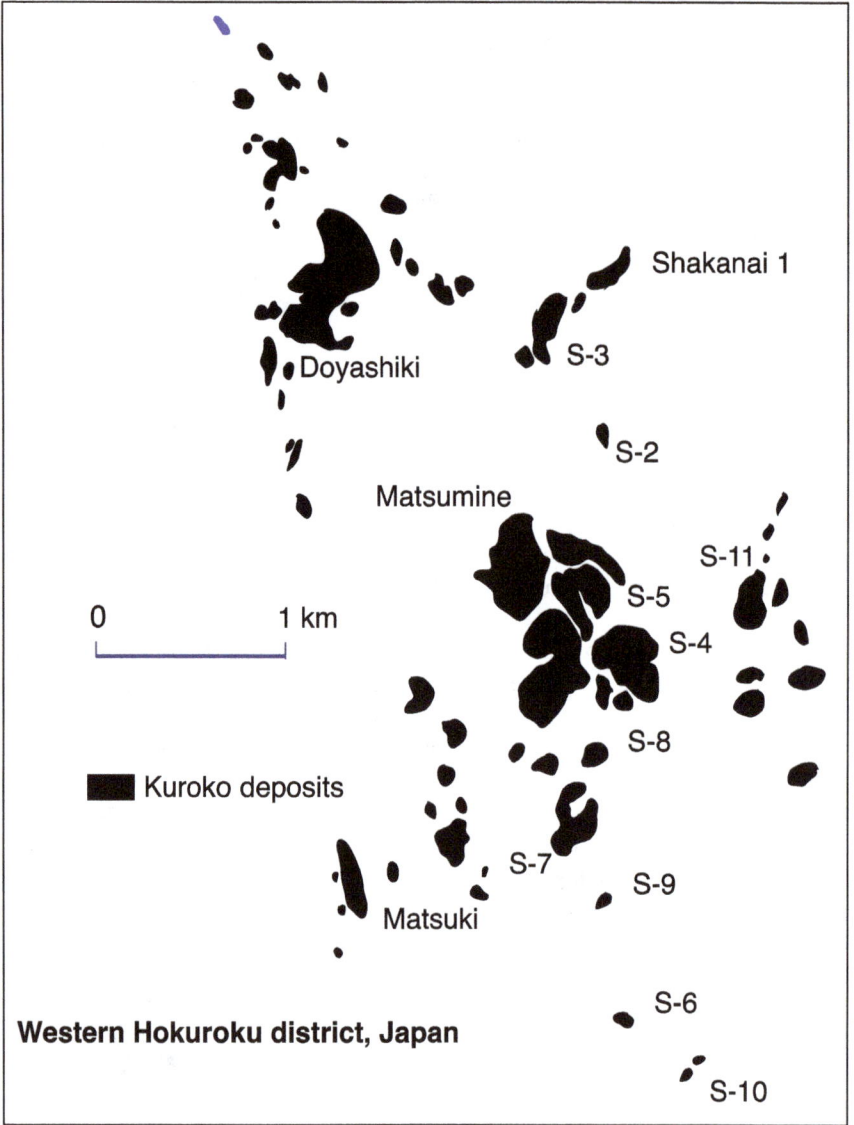

Fig. 22.2 Kuroko volcanogenic massive sulfide deposits of the western part of the Hokuroku district in Northern Japan (after Ohmoto and Takahashi 1983)

in Canada and also contain grades and tonnages of the many mines thus double counting and generating biased metal statistics and frequency distributions of questionable value. There are databases in which spatial rules for combining adjacent deposits have been consistently applied and multiple names have been eliminated (e.g., Mosier et al. 2009). Compilations that use the above sources combined with other sources of data on, for example, volcanic-hosted massive sulfide deposits very likely contain deposits and prospects counted twice (e.g., Patiño-Douce 2016), resulting in statistical analyses that are seriously flawed because the studied population does not represent the target population of mineral resources. Operational rules defining deposits need to account for these map scale effects and for the fact that some deposits have multiple names, mines and separate reported tonnages (Singer 2017).

Mineral occurrences or prospects which are the focus of prospectivity analysis do not qualify as economic mineral deposits because they are typically quite small and incompletely explored. Because number of undiscovered deposits estimates must be defined in a way that is linked to the grade-tonnage or contained metal models, estimates of number of deposits made using models based on such flawed grade-tonnage models must also be a mismatch with the target population.

22.3 Examples of Mismatches in Assessments

Solving the wrong problem due to mismatches of the target population with the studied or estimated population abound in mineral resource assessments. Examples of mismatches include issues of not understanding where the undiscovered resources might exist and estimating something other than mineral deposits that might be economic to mine (De Young and Singer 1981).

In one example, five or more epithermal gold vein deposits were estimated at the 90% level but no grade-and-tonnage model was provided, so the estimated deposits could be any size (Singer and Menzie 2010). To provide critical information to decision-makers, a grade-and-tonnage or contained metal model is key, and the estimated number of deposits that might exist must be from the linked grade-and-tonnage frequency distributions. Estimates of number of undiscovered deposits are completely arbitrary unless tied to a grade-and-tonnage or contained metal model that has been defined in a consistent operational manner.

In an unpublished study, four geoscientists made subjective probabilistic estimates of the number of undiscovered hot-spring mercury deposits in a 1:250,000 scale quadrangle in Alaska. They made independent estimates at the 90th, 50th, and 10th percentiles (Table 22.1). The 10th percentile, for example, is the number of deposits for which there is at least a 10% chance of that number of deposits or more exist.

It was pointed out to participant D that because the number of deposit estimates must be consistent with the grade and tonnage model, his estimates imply that there is more undiscovered mercury in this quadrangle than has been found in the world

Table 22.1 Independent estimates by four scientists of the number of undiscovered hot-spring Hg deposits in a quadrangle in Alaska

Participant	A	B	C	D	
90% chance of at least	1	1	2	9,000	Deposits
50% chance of at least	3	2	4	10,000	Deposits
10% chance of at least	6	6	7	11,000	Deposits

in this deposit type. He responded that he was estimating wisps of cinnabar, not deposits consistent with the grade and tonnage model. In this case, the population considered by participant D did not match the target population. Using a variety of different guidelines such as deposit densities (Singer 2008) for estimates of the number of undiscovered deposits provides a useful crosscheck of assumptions that may have been relied upon and discourages mismatches between target and estimated populations. In these examples of errors in estimating the number of undiscovered deposits, the key is the difference between the understanding of what was being estimated and the population of interest.

In Harris's landmark study (1965), multiple discriminate analysis was used to predict value of mineral production—among the best predictors was geologic cover with a negative value. In a study by Singer (1971), multiple regression was used to predict mineral production and again, cover with a negative value was an important variable. Unlike in petroleum exploration, minerals exploration under cover is a developing technology. Most commonly, mineral exploration under cover results from trying to extend known deposits, that is, additions to reserves. More difficult discovery and higher costs relative to exposed deposits, tend to reduce interest in covered areas. Covered areas tend to be poorly explored and, consequently, deposits under cover tend to be underreported.

In situations where resource assessments are made based on local information, the possibility of solving the wrong problem is high. For example, if the mapped geology were used to predict where and how many undiscovered orogenic gold deposits might in the Bendigo Zone of Victoria Australia, one would conclude that deposits are clustered in space and gold deposits are related to older rocks and covered areas would be worst place to look (Fig. 22.3). Even if we use some modern tools like weights of evidence or neural networks, we would predict no undiscovered deposits under cover. Yet, because geology permissive for the gold deposits is known under cover, and exposed permissive geology is thoroughly explored, most experts would recommend exploration under cover (Lisitin et al. 2007).

Each of these examples demonstrates mismatches of the target population and the studied population. Type III errors in these cases could produce useless or, even worse, misleading assessments.

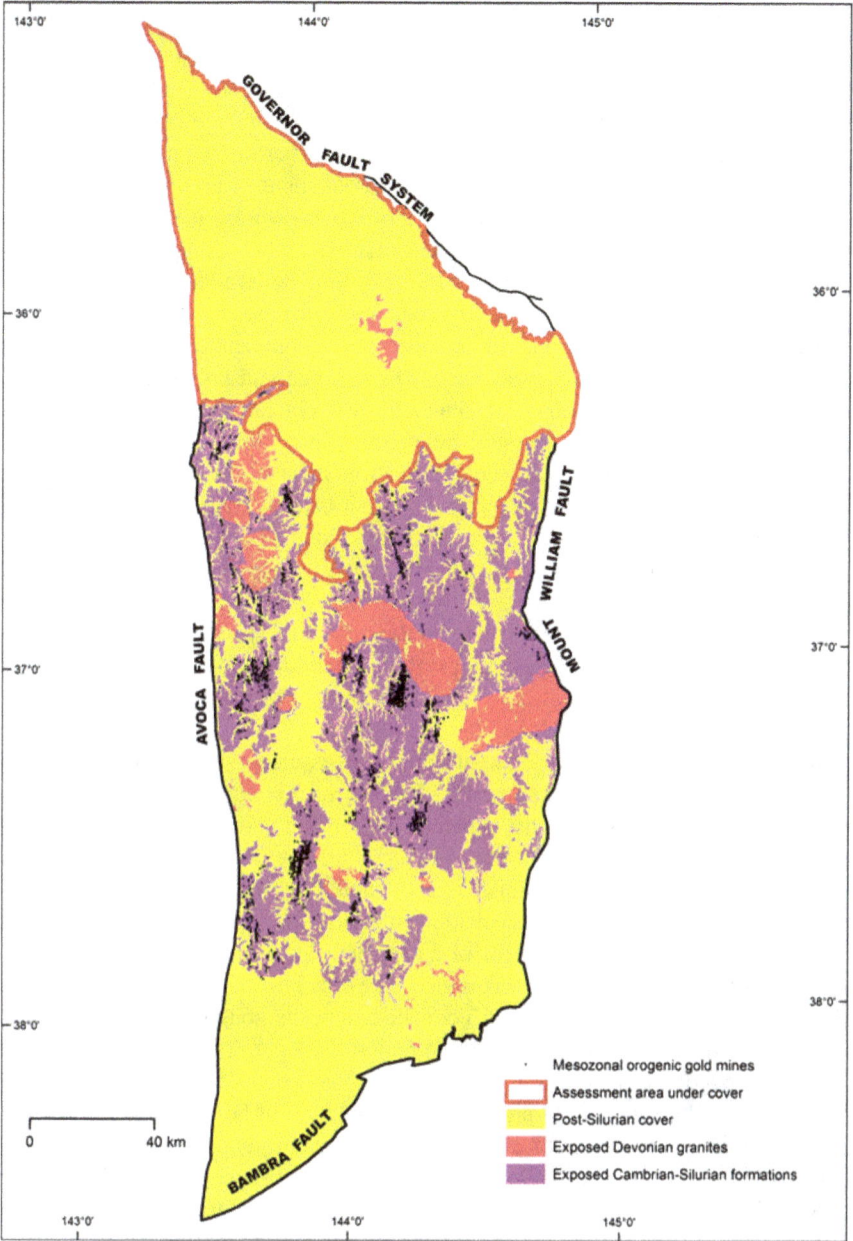

Fig. 22.3 Geology and known orogenic gold deposits (black) in the Bendigo Zone of Victoria, Australia (modified after Lisitsin et al. 2007)

22.4   How to Correct Type III Errors

The problems of mineral resource assessment can only be solved if they are formulated in a way consistent with the decision-maker's language and understanding of the problem. The questions need to be asked: Why perform an assessment? Who is the study being done for and what are the problems they are trying to resolve?

We start with the question of what kinds of issues decision makers are trying to resolve and what types and forms of information would aid in resolving these issues. Unfortunately, the decision-maker may not be available for the needed insight or may not be able to clearly state the information needs. Because the primary purpose of the kinds of assessments recommended here is to help decision-makers determine consequences of economic and policy decisions about tracts of land, regions, countries, or the earth, it is critical that the assessments be unbiased. For example, if the question concerns the long-term supply of a metal, the data used should not contain biased information such as grades and tonnages on multiple versions of the same deposits. These situations require care in compiling data and using sources that report locations, other names of deposits and names of deposits that have been combined with the primary deposit to meet spatial combination rules. A reliable source (e.g., Mosier et al. 2009) has specific information about locations, rules used to combine deposits and specific names that were combined for each deposit. These kinds of data provide a reliable basis for testing statistical distributions of metals in mineral deposits such as the lognormal distribution (Singer 2013).

It is important to recognize that success of assessments depends on the assessments following an integrated approach. This means that no part of the models and methods of estimation have any meaning in isolation. For instance, estimates of number of undiscovered deposits are completely arbitrary unless tied to a grade and tonnage or contained metal model. The goal should be to make explicit the factors that can affect a mineral-related decision so that the decision-maker can clearly see what are the possible consequences of decisions (Singer and Menzie 2010).

To avoid situations where occurrences are the basis of information used to discriminate barren areas from the economic deposits sought, it is necessary to construct models based on the economic deposits sought. Mineral deposit models can be based on data gathered from well-explored deposits of each type from around the world. This would allow the determination of how commonly different attributes and combinations of attributes occur. Quantifying mineral deposit attributes is the necessary and sufficient next step in statistically classifying known deposits by type. Quantified deposit attributes also can provide a firm foundation to identify which observations on geologic and other maps should be effective in delineation of tracts and perhaps identifying sites for detailed exploration. The kind of digital models advocated here would require the recording of both absolute time units and the relative time units of spatially related mineral deposits, rocks, geochemistry, geophysics, and tectonics. The scale of the observations is critical to proper application of such models. This is required to properly apply the models in new geologic settings. Information in these models about the attributes associated with known deposits is necessary but not sufficient to discriminate barren from mineralized environments; quantifying the attributes of barren environments also is necessary for this task. Such digital models could be the foundation for identifying the discriminating functions that could remove many type III errors in assessments.

The exploration department of a major zinc producer found it essential to document a robust decision-making process to maintain internal and investor support (Penney et al. 2004). Zinc deposits from around the world were classed by type, grade, and tonnage models developed for each, cost filters were applied to each, and tracts around the world were delineated where the types could occur (Penney et al. 2004). This study was designed to aid the exploration decision-makers plan the search for economic deposits. Their process was the same as that recommended in three-part assessments (Singer and Menzie 2010), with the exception that they ranked or scored tracts rather than estimating the number of undiscovered deposits.

22.5   Conclusions

Errors of solving the wrong mineral resource problem can make a study's value negative. Type III errors, solving the wrong problem, can be avoided by using care in matching the information needed to solve the decision-maker's problem with information provided in the study. In some cases, we know how to solve the wrong problem but not the real one. It is not uncommon to get rewarded for publishing an answer—not THE answer. With some care and critical thinking in the planning stages, it is possible to provide information useful to decision-makers and to be rewarded for a publication.

References

DeYoung JH Jr, Singer DA (1981) Physical factors that could restrict mineral supply. In: Skinner BJ (ed) 75th anniversary volume. Economic geology, pp 939–954

Harris DP (1965) An application of multivariate statistical analysis to mineral exploration. PhD dissertation, The Pennsylvania State University, 265 pp

Lisitsin V, Olshina A, Moore DH, Willman CE (2007) Assessment of undiscovered mesozonal orogenic gold endowment under cover in the northern part of the Bendigo Zone. GeoScience Victoria Gold Undercover Report 2, Department of Primary Industries, 98 pp. www.dpi.vic.gov.au/minpet/store

Mitroff II, Silvers A (2009) Dirty rotten strategies—how we trick ourselves and others into solving the wrong problems precisely. Stanford University Press, Stanford, 232 pp

Mosier DL, Berger VI, Singer DA (2009) Volcanogenic massive sulfide deposits of the world—database and grade and tonnage models. U.S. Geological Survey Open-file Report 2009-1034. http://pubs.usgs.gov/of/2009/1034/

Ohmoto H, Takahashi T (1983) Submarine calderas and Kuroko genesis. In: Omoto H, Skinner BJ (eds) The Kuroko and related volcanogenic massive sulfide deposits. Econ Geol Monogr 5:38–54

Patiño-Douce AE (2016) Metallic mineral resources in the twenty first century. II. Constraints on future supply. Nat Resour Res 25(1):97–124

Penney SR, Allen RM, Harrisson S, Lees TC, Murphy FC, Norman AR, Roberts PA (2004) A global-scale exploration risk analysis technique to determine the best mineral belts for exploration. Trans Inst Min Metall Sect B. Appl Earth Sci 113:B183–B196

Raiffa H (1968) Decision analysis—introductory lectures on choices under uncertainty. Random House, New York, 309 pp

Singer DA (1971) Multivariate statistical analysis of the unit regional value of mineral resources. Unpublished PhD thesis, The Pennsylvania State University, 210 pp

Singer DA (2017) Future copper resources. Ore Depos Rev 86:271–278. http://dx.doi.org/10.1016/j.oregeorev.2017.02.022

Singer DA (2008) Mineral deposit densities for estimating mineral resources. Math Geosci 40(1):33–46

Singer DA (2013) The lognormal distribution of metal resources in mineral deposits. Ore Geol Rev 55:80–86. http://dx.doi.org/10.1016/j.oregeorev.2013.04.009

Singer DA, Menzie WD (2010) Quantitative mineral resource assessments—an integrated approach. Oxford University Press, New York, 219 pp

The contributors of this book come from diverse backgrounds, making this book a truly international effort. We would like to thank all the contributing authors for lending their expertise to make the book truly unique. They have played a crucial role in the development of this book. Without their invaluable contributions this book wouldn't have been possible. They have made vital efforts to compile up to date information on the varied aspects of this subject to make this book a valuable addition to the collection of many professionals and students.

This book was conceptualized with the vision of imparting up-to-date and integrated information in this field. To ensure the same, a matchless editorial board was set up. Every individual on the board went through rigorous rounds of assessment to prove their worth. After which they invested a large part of their time researching and compiling the most relevant data for our readers.

The editorial board has been involved in producing this book since its inception. They have spent rigorous hours researching and exploring the diverse topics which have resulted in the successful publishing of this book. They have passed on their knowledge of decades through this book. To expedite this challenging task, the publisher supported the team at every step. A small team of assistant editors was also appointed to further simplify the editing procedure and attain best results for the readers.

Apart from the editorial board, the designing team has also invested a significant amount of their time in understanding the subject and creating the most relevant covers. They scrutinized every image to scout for the most suitable representation of the subject and create an appropriate cover for the book.

The publishing team has been an ardent support to the editorial, designing and production team. Their endless efforts to recruit the best for this project, has resulted in the accomplishment of this book. They are a veteran in the field of academics and their pool of knowledge is as vast as their experience in printing. Their expertise and guidance has proved useful at every step. Their uncompromising quality standards have made this book an exceptional effort. Their encouragement from time to time has been an inspiration for everyone.

The publisher and the editorial board hope that this book will prove to be a valuable piece of knowledge for students, practitioners and scholars across the globe.

Index

www.ingramcontent.com/pod-product-compliance
Lightning Source LLC
Chambersburg PA
CBHW050124240326
41458CB00122B/1228